Statistical Mechanics and Thermodynamics of Liquids and Crystals

Statistical Mechanics and Thermodynamics of Liquids and Crystals

Editor

Santi Prestipino

MDPI • Basel • Beijing • Wuhan • Barcelona • Belgrade • Manchester • Tokyo • Cluj • Tianjin

Editor
Santi Prestipino
University of Messina
Italy

Editorial Office
MDPI
St. Alban-Anlage 66
4052 Basel, Switzerland

This is a reprint of articles from the Special Issue published online in the open access journal *Entropy* (ISSN 1099-4300) (available at: https://www.mdpi.com/journal/entropy/special_issues/statistical_mechanics_liquids_crystals).

For citation purposes, cite each article independently as indicated on the article page online and as indicated below:

LastName, A.A.; LastName, B.B.; LastName, C.C. Article Title. *Journal Name* **Year**, *Volume Number*, Page Range.

ISBN 978-3-0365-1551-9 (Hbk)
ISBN 978-3-0365-1552-6 (PDF)

© 2021 by the authors. Articles in this book are Open Access and distributed under the Creative Commons Attribution (CC BY) license, which allows users to download, copy and build upon published articles, as long as the author and publisher are properly credited, which ensures maximum dissemination and a wider impact of our publications.

The book as a whole is distributed by MDPI under the terms and conditions of the Creative Commons license CC BY-NC-ND.

Contents

About the Editor . . . . **vii**

Santi Prestipino
Statistical Mechanics and Thermodynamics of Liquids and Crystals
Reprinted from: *Entropy* **2021**, *23*, 715, doi:10.3390/e23060715 . . . . **1**

Mariano López de Haro, Andrés Santos and Santos B. Yuste
Equation of State of Four- and Five-Dimensional Hard-Hypersphere Mixtures
Reprinted from: *Entropy* **2020**, *22*, 469, doi:10.3390/e22040469 . . . . **5**

Yikun Wei, Pingping Shen, Zhengdao Wang, Hong Liang and Yuehong Qian
Time Evolution Features of Entropy Generation Rate in Turbulent Rayleigh-Bénard Convection with Mixed Insulating and Conducting Boundary Conditions
Reprinted from: *Entropy* **2020**, *22*, 672, doi:10.3390/e22060672 . . . . **23**

Santi Prestipino and Paolo V. Giaquinta
Entropy Multiparticle Correlation Expansion for a Crystal
Reprinted from: *Entropy* **2020**, *22*, 1024, doi:10.3390/e22091024 . . . . **39**

Vera Grishina, Vyacheslav Vikhrenko and Alina Ciach
Structural and Thermodynamic Peculiarities of Core-Shell Particles at Fluid Interfaces from Triangular Lattice Models
Reprinted from: *Entropy* **2020**, *22*, 1215, doi:10.3390/e22111215 . . . . **61**

Santi Prestipino
Ultracold Bosons on a Regular Spherical Mesh
Reprinted from: *Entropy* **2020**, *22*, 1289, doi:10.3390/e22111289 . . . . **79**

Luis M. Sesé
Real Space Triplets in Quantum Condensed Matter: Numerical Experiments Using Path Integrals, Closures, and Hard Spheres
Reprinted from: *Entropy* **2020**, *22*, 1338, doi:10.3390/e22121338 . . . . **99**

Sergii D. Kaim
The Molecular Theory of Liquid Nanodroplets Energetics in Aerosols
Reprinted from: *Entropy* **2021**, *23*, 13, doi:10.3390/e23010013 . . . . **127**

Nkosinathi Dlamini, Santi Prestipino, Giuseppe Pellicane
Self-Assembled Structures of Colloidal Dimers and Disks on a Spherical Surface
Reprinted from: *Entropy* **2021**, *23*, 585, doi:10.3390/e23050585 . . . . **153**

About the Editor

Santi Prestipino (Professor), born in 1966, is a theoretical and computational physicist at the Università degli Studi di Messina (Italy). His long-term research topics include the thermodynamics and statistical mechanics of liquids, crystals, and their interfaces. In the last few years, he has been involved in problems related with crystal nucleation, colloidal self-assembly, and the phases of ultracold bosonic gases. He is on the Editorial Board of *Entropy* (MDPI) and *Foundations* (MDPI).

Editorial

Statistical Mechanics and Thermodynamics of Liquids and Crystals

Santi Prestipino

Dipartimento di Scienze Matematiche ed Informatiche, Scienze Fisiche e Scienze della Terra, Università degli Studi di Messina, Viale F. Stagno d'Alcontres 31, 98166 Messina, Italy; sprestipino@unime.it

Citation: Prestipino, S. Statistical Mechanics and Thermodynamics of Liquids and Crystals. *Entropy* **2021**, *23*, 715. https://doi.org/10.3390/e23060715

Academic Editor: Antonio M. Scarfone

Received: 2 June 2021
Accepted: 3 June 2021
Published: 4 June 2021

Publisher's Note: MDPI stays neutral with regard to jurisdictional claims in published maps and institutional affiliations.

Copyright: © 2021 by the author. Licensee MDPI, Basel, Switzerland. This article is an open access article distributed under the terms and conditions of the Creative Commons Attribution (CC BY) license (https://creativecommons.org/licenses/by/4.0/).

Thermodynamic phases are the most prominent manifestation of emergent behavior. Among them, crystals and liquids traditionally epitomize the antagonistic concepts of order and disorder (i.e., the presence or absence of a symmetry). According to common wisdom, the competition for stability between solid and liquid reflects the struggle between the opposite tendencies of energy minimization and entropy maximization, which is regulated by temperature. It goes without saying that the previous statement is in fact too simplistic, since in equilibrium the guiding principle is rather the minimization of the thermodynamic potential (maximization of the Massieu function) appropriate to the given control parameters. For instance, crystallization sometimes occurs with the purpose to maximize entropy (as for hard spheres under pressure). A further example is provided by superfluid, which can be one of the phases of minimum energy/enthalpy at zero temperature.

The relevance of the solid/liquid dichotomy for statistical physics cannot simply be overstated. It was under the pressure of accurately locating the solid–liquid transition in simple-fluid models that "exact" free-energy methods were initially developed [1,2]. Since then, variations on these methods have been employed to compute the phase diagram of many complex fluids, such as liquid crystals [3], cluster crystals [4], and fluids of patchy particles [5]. Generally speaking, the full control of phase behavior can help in the synthesis of artificial materials with the desired specifications.

Turning to theory, while the exact determination of the partition function for a non-trivial system with solid, liquid, and vapor phases will probably never be accomplished, there are nevertheless variational treatments of the solid–liquid transition (density functional theories) that have by now reached a high degree of sophistication (see, e.g., Refs. [6,7]). However, a schematic phase diagram with the standard three phases can be obtained with less effort, see for instance the mean-field analysis of the Potts lattice gas [8].

It is usually thought that (classical) liquid is a unique phase, while crystals are a multitude. In fact, this is only partially true, since liquids composed of hydrogen-bonded molecules exhibit a number of so-called water-like "anomalies" that make them different in many respects from conventional (rare-gas) fluids (see, e.g., [9]). The relationship between such anomalies and solid polymorphism/polyamorphism is an important topic of statistical physics. Phase diagrams with many solid polymorphs are the rule for simple substances (e.g., Na [10]) under huge pressure. Here, exact free-energy methods face a serious limitation, since the crystalline structures—Bravais and non-Bravais—being potentially relevant are countless. Metadynamics [11] and evolutionary algorithms [12] are a possible way out, since they do not involve any assumption on the topology of the energy landscape.

Condensed-matter physics also features a wide variety of phases with mixed solid and liquid characteristics. Hexatic fluids [13], liquid-crystal smectics [14], crystalline membranes [15], and quantum supersolids [16] are just a few examples of hybrid states of aggregation, with others yet to be discovered. In the last few years, a new category of systems, i.e., self-assembling materials, has fallen under the scrutiny of the statistical-physics community (see, e.g., [17]). Here, some kind of order at the mesoscopic scale spontaneously emerges from the interaction between simple microscopic units. For these systems, the precise interplay between thermodynamics and kinetics in the onset of aggregates is still being worked out. As we move forward in the bottom-up investigation of biological matter, the concerted role of energy and entropy in the formation of structures with hierarchical order will be made more clear.

This Special Issue collects articles published between April 2020 and May 2021, highlighting novel results in the application of statistical thermodynamics to liquids and crystals.

In the first of these articles [18], the focus is on the virial equation of state for mixtures of hard hyperspheres. While such systems obviously have no counterpart in the real world, they still represent a useful playground where to explore the effectiveness of approximations routinely employed in three dimensions. If geometric considerations play a leading role in the crossover from three to five dimensions, then simple analytic extensions of approximations that proved successful in lower dimensions would provide accurate equations of state for hard hyperspheres in the fluid phase. It turns out that the sole requirement of reproducing the exact second and third virial coefficients yields, in four and five dimensions, approximate equations of state of overall good quality in the comparison with computer-simulation data.

Moderately dense particles driven far from equilibrium are much harder to attack theoretically. In this case, numerical simulation is the only method to assess the accuracy of fluid dynamic equations. The second article in this Special Issue [19] is an attempt to investigate turbulent convection using the Boussinesq approximation—accounting for the variation of density with temperature only in the buoyancy term of the Navier–Stokes equation. The approach of numerical simulation allows one to analyze both heat and mass flow under a variety of boundary conditions, with potential applications in oceanography, geophysics, astrophysics, and industry.

A well-established approach to the entropy of a simple fluid is the so-called multiparticle correlation expansion (MPCE), expressing the statistical entropy as a sum of contributions from increasingly large numbers of particles. Upon truncating the MPCE after the two-body term S_2, one has an estimate of the exact entropy that turns out to be accurate right at the point of transition into a crystal, leading to a freezing criterion [20] that has had a certain success in the past. After reviewing the history of the entropy MPCE, Ref. [21] inquires into the possibility of formulating an analogous entropic criterion for melting, given that a MPCE formally holds also for the entropy of a crystal. However, the computation of S_2 proves to be a formidable task even for a crystal of hard disks, thus dampening the enthusiasm for any melting criterion based on the numerical evaluation of the two-body entropy.

Reference [22] investigates pattern formation in a two-dimensional lattice gas system. Lattice gases allow for an exact thermodynamic analysis at zero temperature, since all possible ground states can be enumerated and compared with each other in terms of enthalpy. Such studies can be helpful to design and control the functionalization of colloidal particles with polymers.

Hard-core bosons are the quantum counterpart of lattice-gas particles. Lattice systems of bosonic particles provide models where the competition between itinerant and localized quantum states can be examined in full detail. The prototype of all such models is the celebrated Bose-Hubbard model, describing the behavior of ultracold bosonic atoms trapped in an optical lattice. Upon increasing the intensity of laser light, the confining potential gets deeper until a transition occurs from superfluid to Mott insulator (i.e., a normal cluster solid) [23,24]. The Bose–Hubbard model and its variants provide an ideal setting for exploring strong-correlation effects in quantum systems, which now are also studied for bosons on the nodes of a spherical mesh [25,26]. However weird this geometry may seem, traps located at the vertices of a polyhedron can be fabricated with optical tweezers and loaded with Rydberg atoms [27]. In particular, a system of bosons in a cubic mesh [25] offers the opportunity to assess the validity of mean-field theory, as well as to uncover the manifestations of superfluidity in a small quantum system.

Spatial correlations between triplets of particles in a fluid or solid are notoriously difficult to investigate in full extent, owing to heavy CPU-time and memory requirements. Yet, triplet correlations convey crucial information for any statistical analysis aiming to go beyond the traditional pairwise approximation. Reference [28] investigates triplet correlations in a fluid of quantum hard spheres, as well as in two crystalline phases of the same system. Using path-integral Monte Carlo simulations, the author of [28] delves into the accuracy of a few closure relations expressing the triplet distribution function in terms of two-body terms, eventually identifying a combination of closures that performs well in rather disparate conditions.

The emergency caused by the COVID-19 pandemic has boosted a large amount of research activities in the last year with the purpose of clarifying the many open questions that arise in connection with the transmission of the infection. Our Special Issue too contains an article on the problem [29], about diffusion in the air of liquid nanodroplets containing the infective agent. Using a molecular theory, the author of [29] derives an effective Hamiltonian for gas atoms and liquid droplets which accounts for the interaction and correlation effects induced by the granular structure of the droplets. Similar theoretical studies may be viewed as complementary to atomistic simulations, which are obviously much more computationally demanding.

The last article in this Special Issue [30] deals with a numerical investigation of the self-assembling behavior of a system made up of asymmetric dimers and marbles ("disks") confined in a spherical surface, as is realized by, e.g., a mixture of colloidal particles spread over the surface of an oil droplet. In the model, the formation of disk aggregates is triggered by a short-range attraction between the disk and one of the monomers. For low disk compositions, only small clusters are found, while for higher composition values, we observe the onset of long flexible chains which, at sufficiently high density, give origin to an intricate network on the sphere. When disks are much larger than dimers, square-ordered patches are formed instead, similar to the truncated triangular crystals of polystyrene spheres growing on the inside walls of water droplets [31], and in striking contrast to the spanning triangular crystal, punctuated by islands of defects, that is promoted by entropy alone in dense hard disks on a sphere [32].

It is our hope that this Special Issue leaves the reader with the impression that the field of liquids and crystals is a vivid research area, full of problems still waiting for solution, and open to surprises.

Funding: This research received no external funding.

Conflicts of Interest: The author declares no conflict of interest.

References

1. Widom, B. Some Topics in the Theory of Fluids. *J. Chem. Phys.* **1963**, *39*, 2808–2812. [CrossRef]
2. Frenkel, D.; Ladd, A.J.C. New Monte Carlo method to compute the free energy of arbitrary solids. Application to the fcc and hcp phases of hard spheres. *J. Chem. Phys.* **1984**, *81*, 3188–3193. [CrossRef]
3. Bolhuis, P.; Frenkel, D. Tracing the phase boundaries of hard spherocylinders. *J. Chem. Phys.* **1997**, *106*, 666–687. [CrossRef]
4. Zhang, K.; Charbonneau, P.; Mladek, B.M. Reentrant and Isostructural Transitions in a Cluster-Crystal Former. *Phys. Rev. Lett.* **2010**, *105*, 245701. [CrossRef]
5. Doppelbauer, G.; Noya, E.G.; Bianchi, E.; Kahl, G. Self-assembly scenarios of patchy colloidal particles. *Soft Matter.* **2012**, *8*, 7768–7772. [CrossRef]
6. Tarazona, P. Density Functional for Hard Sphere Crystals: A Fundamental Measure Approach. *Phys. Rev. Lett.* **2000**, *84*, 694–697. [CrossRef]
7. Warshavsky, V.B.; Song, X. Calculations of free energies in liquid and solid phases: Fundamental measure density-functional approach. *Phys. Rev. E* **2004**, *69*, 061113. [CrossRef] [PubMed]
8. Shih, W.-H.; Ebner, C.; Stroud, D. Potts lattice-gas model for the solid–liquid interfacial tensions of simple fluids. *Phys. Rev. B* **1986**, *34*, 1811–1814. [CrossRef] [PubMed]
9. Russo, J.; Akahane, K.; Tanaka, H. Water-like anomalies as a function of tetrahedrality. *Proc. Natl. Acad. Sci. USA* **2018**, *115*, 3333–3341. [CrossRef] [PubMed]
10. Eshet, H.; Khaliullin, R.Z.; Kühne, T.D.; Behler, J.; Parrinello, M. Microscopic Origins of the Anomalous Melting Behavior of Sodium under High Pressure. *Phys. Rev. Lett.* **2012**, *108*, 115701. [CrossRef]

11. Martoňák, R.; Laio, A.; Parrinello, M. Predicting Crystal Structures: The Parrinello-Rahman Method Revisited. *Phys. Rev. Lett.* **2003**, *90*, 075503. [CrossRef] [PubMed]
12. Oganov, A.R.; Glass, C.W. Crystal structure prediction using ab initio evolutionary techniques: Principles and applications. *J. Chem. Phys.* **2006**, *124*, 244704. [CrossRef] [PubMed]
13. Prestipino, S.; Saija, F. Hexatic phase and cluster crystals of two-dimensional GEM4 spheres. *J. Chem. Phys.* **2014**, *141*, 184502. [CrossRef] [PubMed]
14. Frenkel, D.; Lekkerkerker, H.N.W.; Stroobants, A. Thermodynamic stability of a smectic phase in a system of hard rods. *Nature* **1988**, *332*, 822–823. [CrossRef]
15. Seung, H.S.; Nelson, D.R. Defects in flexible membranes with crystalline order. *Phys. Rev. A* **1988**, *38*, 1005–1018. [CrossRef]
16. Boninsegni, M.; Prokof'ev, N.V. Supersolids: What and where are they? *Rev. Mod. Phys.* **2012**, *84*, 759–776. [CrossRef]
17. Sciortino, F. Entropy in self-assembly. *Riv. Nuovo Cim.* **2019**, *42*, 511–548.
18. López de Haro, M.; Santos, A.; Yuste, S.B. Equation of State of Four- and Five-Dimensional Hard-Hypersphere Mixtures. *Entropy* **2020**, *22*, 469. [CrossRef] [PubMed]
19. Wei, Y.; Shen, P.; Wang, Z.; Liang, H.; Qian, Y. Time Evolution Features of Entropy Generation Rate in Turbulent Rayleigh-Bénard Convection with Mixed Insulating and Conducting Boundary Conditions. *Entropy* **2020**, *22*, 672. [CrossRef]
20. Giaquinta, P.V.; Giunta, G. About entropy and correlations in a fluid of hard spheres. *Physica A* **1992**, *187*, 145–158. [CrossRef]
21. Prestipino, S.; Giaquinta, P.V. Entropy Multiparticle Correlation Expansion for a Crystal. *Entropy* **2020**, *22*, 1024. [CrossRef]
22. Grishina, V.; Vikhrenko, V.; Ciach, A. Structural and Thermodynamic Peculiarities of Core-Shell Particles at Fluid Interfaces from Triangular Lattice Models. *Entropy* **2020**, *22*, 1215. [CrossRef]
23. Jaksch, D.; Bruder, C.; Cirac, J.I.; Gardiner, C.W.; Zoller, P. Cold Bosonic Atoms in Optical Lattices. *Phys. Rev. Lett.* **1998**, *81*, 3108–3111. [CrossRef]
24. Greiner, M.; Mandel, O.; Esslinger, T.; Hänsch, T.W.; Bloch, I. Quantum phase transition from a superfluid to a Mott insulator in a gas of ultracold atoms. *Nature* **2002**, *415*, 39–44. [CrossRef]
25. Prestipino, S. Ultracold Bosons on a Regular Spherical Mesh. *Entropy* **2020**, *22*, 1289. [CrossRef] [PubMed]
26. Prestipino, S. Bose-Hubbard model on polyhedral graphs. *Phys. Rev. A* **2021**, *103*, 033313. [CrossRef]
27. Browaeys, A.; Lahaye, T. Many-body physics with individually controlled Rydberg atoms. *Nat. Phys.* **2020**, *16*, 132–142. [CrossRef]
28. Sesé, L.M. Real Space Triplets in Quantum Condensed Matter: Numerical Experiments Using Path Integrals, Closures, and Hard Spheres. *Entropy* **2020**, *22*, 1338. [CrossRef]
29. Kaim, S.D. The Molecular Theory of Liquid Nanodroplets Energetics in Aerosols. *Entropy* **2021**, *23*, 13. [CrossRef]
30. Dlamini, N.; Prestipino, S.; Pellicane, G. Self-Assembled Structures of Colloidal Dimers and Disks on a Spherical Surface. *Entropy* **2021**, *23*, 585. [CrossRef] [PubMed]
31. Meng, G.; Paulose, J.; Nelson, D.R.; Manoharan, V.N. Elastic Instability of a Crystal Growing on a Curved Surface. *Science* **2014**, *343*, 634–637. [CrossRef] [PubMed]
32. Prestipino, S.; Ferrario, M.; Giaquinta, P.V. Statistical geometry of hard particles on a sphere: Analysis of defects at high density. *Physica A* **1993**, *201*, 649–665.

Article

Equation of State of Four- and Five-Dimensional Hard-Hypersphere Mixtures

Mariano López de Haro [1], Andrés Santos [2],* and Santos B. Yuste [2]

[1] Instituto de Energías Renovables, Universidad Nacional Autónoma de México (U.N.A.M.), Temixco, Morelos 62580, Mexico; malopez@unam.mx
[2] Departamento de Física and Instituto de Computación Científica Avanzada (ICCAEx), Universidad de Extremadura, E-06006 Badajoz, Spain; santos@unex.es
* Correspondence: andres@unex.es; Tel.: +34-924-289-651

Received: 24 March 2020; Accepted: 16 April 2020; Published: 20 April 2020

Abstract: New proposals for the equation of state of four- and five-dimensional hard-hypersphere mixtures in terms of the equation of state of the corresponding monocomponent hard-hypersphere fluid are introduced. Such proposals (which are constructed in such a way so as to yield the exact third virial coefficient) extend, on the one hand, recent similar formulations for hard-disk and (three-dimensional) hard-sphere mixtures and, on the other hand, two of our previous proposals also linking the mixture equation of state and the one of the monocomponent fluid but unable to reproduce the exact third virial coefficient. The old and new proposals are tested by comparison with published molecular dynamics and Monte Carlo simulation results and their relative merit is evaluated.

Keywords: equation of state; hard hyperspheres; fluid mixtures

1. Introduction

The interest in studying systems of d-dimensional hard spheres has been present for many decades and still continues to stimulate intensive research [1–96]. This interest is based on the versatility of such systems that allows one to gain insight into, among other things, the equilibrium and dynamical properties of simple fluids, colloids, granular matter, and glasses with which they share similar phenomenology. For instance, it is well known that all d-dimensional hard-sphere systems undergo a fluid-solid phase transition which occurs at smaller packing fractions as the spatial dimension is increased. This implies that mean-field-like descriptions of this transition become mathematically simpler and more accurate as one increases the number of dimensions. Additionally, in the limit of infinite dimension one may even derive analytical results for the thermodynamics, structure, and phase transitions of such hypersphere fluids [1–13]. In particular, the equation of state (EOS) truncated at the level of the second virial coefficient becomes exact in this limit [8].

While of course real experiments cannot be performed in these systems, they are amenable to computer simulations and theoretical developments. Many aspects concerning hard hyperspheres have been already dealt with, such as thermodynamic and structural properties [13–67], virial coefficients [67–80], and disordered packings [52,81–91] or glassy behavior [12,81,82,92]. Nevertheless, due to the fact that (except in the infinite dimensional case) no exact analytical results are available, efforts to clarify or reinforce theoretical developments are worth pursuing. In the case of mixtures of hard hyperspheres this is particularly important since, comparatively speaking, the literature pertaining to them is not very abundant. To the best of our knowledge, the first paper reporting an (approximate) EOS for additive binary hard-hypersphere fluid mixtures is the one by González et al. [28], in which they used the overlap volume approach. What they did was to compute the partial direct correlation functions through an interpolation between the exact low-density and the Percus–Yevick high-density behavior of such functions to produce a Carnahan–Starling-like EOS which they subsequently compared with the (very few then) available simulation data for additive hard-disk mixtures. A few years later, we [32,48] proposed an ansatz for the contact values of the partial radial distribution functions complying with some exact limiting conditions to derive an EOS (henceforth denoted with the label "e1") of a multicomponent d-dimensional hard-sphere fluid in terms of the one of the single monocomponent system. To our knowledge, the first simulation results for the structural and thermodynamic properties of additive hard-hypersphere mixtures were obtained via molecular dynamics (MD) for a few binary mixtures in four and five spatial dimensions by González-Melchor et al. [36], later confirmed by Monte Carlo (MC) computations by Bishop and Whitlock [41]. The comparison between such simulation results and our e1 EOS [32] led to very reasonable agreement. Later, we proposed a closely related EOS (henceforth denoted with the label "e2") stemming from additional exact limiting conditions applied to the contact values of the partial radial distribution functions [37,48]. A limitation of these proposals is that, except in the three-dimensional case, they are unable to yield the exact third virial coefficient. As shown below, extensions of these EOS (denoted as "ē1" and "ē2") complying with the requirement that the third virial coefficient computed from them is the exact one, may be introduced with little difficulty. More recently, we have developed yet another approximate EOS (henceforth denoted with the label "sp") for d-dimensional hard-sphere fluid mixtures [63,64,93], and newer simulation results for hard hypersphere mixtures have also been obtained [57–59]. It is the aim of this paper to carry out a comparison between available simulation data for binary additive four- and five-dimensional hypersphere fluid mixtures and our theoretical proposals.

The paper is organized as follows. In order to make it self-contained, in Section 2 we provide a brief outline of the approaches we have followed to link the EOS of a polydisperse d-dimensional hard-sphere mixture and that of the corresponding monocomponent system. Section 3 presents the specific cases of four and five spatial dimensions, the choice of the EOS of the monocomponent system to complete the mapping, and the comparison with the simulation data. We close the paper in Section 4 with a discussion of the results and some concluding remarks.

2. Mappings between the Equation of State of the Polydisperse Mixture and That of the Monocomponent System

Let us begin by considering a mixture of additive hard spheres in d dimensions with an arbitrary number s of components. This number s may even be infinite, i.e., the system may also be a polydisperse mixture with a continuous size distribution. The additive hard core of the interaction between a sphere of species i and a sphere of species j is $\sigma_{ij} = \frac{1}{2}(\sigma_i + \sigma_j)$, where the diameter of a sphere of species i is $\sigma_{ii} = \sigma_i$. Let the number density of the mixture be ρ and the mole fraction of species i be $x_i = \rho_i/\rho$, where ρ_i is the number density of species i. In terms of these quantities, the packing fraction is given by $\eta = v_d \rho M_d$, where $v_d = (\pi/4)^{d/2}/\Gamma(1+d/2)$ is the volume of a d-dimensional sphere of unit diameter, $\Gamma(\cdot)$ is the Gamma function, and $M_n \equiv \langle \sigma^n \rangle = \sum_{i=1}^{s} x_i \sigma_i^n$ denotes the nth moment of the diameter distribution.

Unfortunately, no exact explicit EOS for a fluid mixture of d-dimensional hard spheres is available. The (formal) virial expression for such EOS involves only the contact values $g_{ij}(\sigma_{ij}^+)$ of the radial distribution functions $g_{ij}(r)$, where r is the distance, namely

$$Z(\eta) = 1 + \frac{2^{d-1}}{M_d} \eta \sum_{i,j=1}^{s} x_i x_j \sigma_{ij}^d g_{ij}(\sigma_{ij}^+), \qquad (1)$$

where $Z = p/\rho k_B T$ is the compressibility factor of the mixture, p being the pressure, k_B the Boltzmann constant, and T the absolute temperature. Hence, a useful way to obtain approximate expressions for the EOS of the mixture is to propose or derive approximate expressions for the contact values $g_{ij}(\sigma_{ij}^+)$. We have already followed this route and the outcome is briefly described in Sections 2.1 and 2.2. More details may be found in Ref. [48] and references therein.

2.1. The e1 Approximation

The basic assumption is that, at a given packing fraction η, the dependence of $g_{ij}(\sigma_{ij}^+)$ on the sets of $\{\sigma_k\}$ and $\{x_k\}$ takes place *only* through the scaled quantity

$$z_{ij} \equiv \frac{\sigma_i \sigma_j}{\sigma_{ij}} \frac{M_{d-1}}{M_d}, \qquad (2)$$

which we express as

$$g_{ij}(\sigma_{ij}^+) = \mathcal{G}(\eta, z_{ij}), \qquad (3)$$

where the function $\mathcal{G}(\eta, z)$ is *universal*, i.e., it is a common function for all the pairs (i,j), regardless of the composition and number of components of the mixture. Next, making use of some consistency conditions, we have derived two approximate expressions for the EOS of the mixture. The first one, labeled "e1," indicating that (i) the contact values $g_{ij}(\sigma_{ij}^+)$ used are an *extension* of the monocomponent fluid contact value $g_s \equiv g(\sigma^+)$ and that (ii) $\mathcal{G}(\eta, z)$ is a *linear* polynomial in z, leads to an EOS that exhibits an excellent agreement with simulations in 2, 3, 4, and 5 dimensions, provided that an accurate g_s is used as input [32,36,57,59,67]. This EOS may be written as

$$Z_{e1}(\eta) = 1 + \frac{\eta}{1-\eta} 2^{d-1}(\Omega_0 - \Omega_1) + [Z_s(\eta) - 1]\Omega_1, \qquad (4)$$

where the coefficients Ω_m depend only on the composition of the mixture and are defined by

$$\Omega_m = 2^{-(d-m)} \frac{M_{d-1}^m}{M_d^{m+1}} \sum_{n=0}^{d-m} \binom{d-m}{n} M_{n+m} M_{d-n}. \qquad (5)$$

It is interesting to point out that from Equation (4) one may write the virial coefficients of the mixture B_n, defined by

$$Z(\rho) = 1 + \sum_{n=1}^{\infty} B_{n+1}\rho^n, \qquad (6)$$

in terms of the (reduced) virial coefficients of the single component fluid b_n defined by

$$Z_s(\eta) = 1 + \sum_{n=1}^{\infty} b_{n+1}\eta^n. \qquad (7)$$

The result is

$$\bar{B}_n^{e1} = \Omega_1 b_n + 2^{d-1}(\Omega_0 - \Omega_1), \qquad (8)$$

where $\bar{B}_n \equiv B_n/(v_d M_d)^{n-1}$ are reduced virial coefficients. Since $b_2 = 2^{d-1}$, Equation (8) yields the *exact* second virial coefficient [63]

$$\bar{B}_2 = 2^{d-1}\Omega_0. \qquad (9)$$

In general, however, $\bar{B}_n^{e1}$ with $n \geq 3$ are only approximate. In particular,

$$\bar{B}_3^{e1} = 1 + \left(\frac{b_3}{4} + 2\right)\frac{M_1 M_3}{M_4} + 3\frac{M_2^2}{M_4} + \left(\frac{3b_3}{4} - 6\right)\frac{M_2 M_3^2}{M_4^2}, \quad (d=4), \qquad (10a)$$

$$\bar{B}_3^{e1} = 1 + \frac{65}{4}\frac{M_1 M_4}{M_5} + 10\frac{M_2 M_3}{M_5} + 45\frac{M_2 M_4^2}{M_5^2} + \frac{135}{4}\frac{M_3^2 M_4}{M_5^2}, \quad (d=5). \qquad (10b)$$

In Equation (10a),

$$b_3 = 64\left(\frac{4}{3} - \frac{3\sqrt{3}}{2\pi}\right), \quad (d=4), \qquad (11)$$

is the reduced third virial coefficient of a monocomponent four-dimensional fluid, while in Equation (10b) we have taken into account that $b_3 = 106$ if $d = 5$.

It is interesting to note that, by eliminating Ω_0 and Ω_1 in favor of $\bar{B}_2$ and $\bar{B}_3^{e1}$, Equation (4) can be rewritten as

$$Z_{e1}(\eta) = 1 + \frac{\eta}{1-\eta} \frac{b_3 \bar{B}_2 - b_2 \bar{B}_3^{e1}}{b_3 - b_2} + [Z_s(\eta) - 1] \frac{\bar{B}_3^{e1} - \bar{B}_2}{b_3 - b_2}. \tag{12}$$

2.2. The e2 Approximation

The second approximation, labeled "e2," similarly indicates that (i) the resulting contact values represent an *extension* of the single component contact value g_s and that (ii) $\mathcal{G}(\eta, z)$ is a *quadratic* polynomial in z. In this case, one also gets a closed expression for the compressibility factor in terms of the packing fraction η and the first few moments M_n, $n \leq d$. Such an expression is

$$Z_{e2}(\eta) = Z_{e1}(\eta) - (\Omega_2 - \Omega_1) \left[Z_s(\eta) \left(1 - 2^{d-2}\eta\right) - 1 - 2^{d-2} \frac{\eta}{1-\eta} \right]. \tag{13}$$

The associated (reduced) virial coefficients are

$$\bar{B}_n^{e2} = \bar{B}_n^{e1} - (\Omega_2 - \Omega_1) \left[b_n - 2^{d-2} (1 + b_{n-1}) \right]. \tag{14}$$

Again, since $b_1 = 1$ and $b_2 = 2^{d-1}$, the exact second virial coefficient, Equation (9), is recovered for any dimensionality. Additionally, in the case of spheres ($d = 3$), $b_3 = 10$ and thus $\bar{B}_3^{e1} = \bar{B}_3^{e2} = 4\Omega_0 + 6\Omega_1$, which is the exact result for that dimensionality. In the cases of $d = 4$ and $d = 5$, one has

$$\bar{B}_3^{e2} = 1 + \left(\frac{b_3}{2} - 7\right) \frac{M_1 M_3}{M_4} + 3 \frac{M_2^2}{M_4} + (b_3 - 15) \frac{M_2 M_3^2}{M_4^2} + \left(18 - \frac{b_3}{2}\right) \frac{M_3^4}{M_4^3}, \quad (d = 4), \tag{15a}$$

$$\bar{B}_3^{e2} = 1 + \frac{25}{2} \frac{M_1 M_4}{M_5} + 10 \frac{M_2 M_3}{M_5} + \frac{75}{2} \frac{M_2 M_4^2}{M_5^2} + \frac{45}{2} \frac{M_3^2 M_4}{M_5^2} + \frac{45}{2} \frac{M_3 M_4^3}{M_5^3}, \quad (d = 5). \tag{15b}$$

It is also worthwhile noting that $\Omega_1 = \Omega_2$ in the case of disks ($d = 2$) and thus $Z_{e1}(\eta) = Z_{e2}(\eta)$ for those systems.

2.3. Exact Third Virial Coefficient. Modified Versions of the e1 and e2 Approximations

As said above, both $\bar{B}_3^{e1}$ and $\bar{B}_3^{e2}$ differ from the exact third virial coefficient, except in the three-dimensional case ($d = 3$). The exact expression is [63]

$$\bar{B}_3 = \frac{1}{M_d^2} \sum_{i,j,k=1}^{s} x_i x_j x_k \widehat{B}_{ijk}, \tag{16a}$$

$$\widehat{B}_{ijk} = \frac{d^2}{3} 2^{5d/2-1} \Gamma(d/2) \left(\sigma_{ij}\sigma_{ik}\sigma_{jk}\right)^{d/2} \int_0^\infty \frac{d\kappa}{\kappa^{1+d/2}} J_{d/2}(\kappa\sigma_{ij}) J_{d/2}(\kappa\sigma_{ik}) J_{d/2}(\kappa\sigma_{jk}), \tag{16b}$$

where $J_n(\cdot)$ is the Bessel function of the first kind of order n.

For odd dimensionality, it turns out that the composition-independent coefficients $\widehat{B}_{ijk}$ have a polynomial dependence on σ_i, σ_j, and σ_k. As a consequence, the third virial coefficient $\bar{B}_3$ can be expressed in terms of moments M_n with $1 \leq n \leq d$. In particular [63],

$$\bar{B}_3 = 1 + 10\frac{M_1 M_4}{M_5} + 20\frac{M_2 M_3}{M_5} + 25\frac{M_2 M_4^2}{M_5^2} + 50\frac{M_3^2 M_4}{M_5^2}, \quad (d = 5). \tag{17}$$

On the other hand, for even dimensionality the dependence of $\widehat{B}_{ijk}$ on σ_i, σ_j, and σ_k is more complex than polynomial. In particular, for a binary mixture ($s=2$) with $d=4$ one has

$$\widehat{B}_{111} = b_3 \sigma_1^8, \quad (d=4), \tag{18a}$$

$$\widehat{B}_{112} = \sigma_1^8 \frac{16(1+q)^4}{3}\left[1 - \frac{1}{8\pi}(1-q)(3+q)(5+2q+q^2)\arcsin\frac{1}{1+q} - \frac{\sqrt{q(2+q)}}{24\pi(1+q)^4}(45 + 138q \right.$$
$$\left. + 113q^2 + 68q^3 + 47q^4 + 18q^5 + 3q^6)\right], \quad (d=4), \tag{18b}$$

where $q \equiv \sigma_2/\sigma_1$ is the size ratio. The expressions for $\widehat{B}_{222}$ and $\widehat{B}_{122}$ can be obtained from Equations (18a) and (18b), respectively, by the replacements $\sigma_1 \to \sigma_2$, $q \to q^{-1}$.

Figure 1 displays the size-ratio dependence of the exact second and third virial coefficients for three representative binary compositions of four- and five-dimensional systems. The degree of bidispersity of a certain binary mixture can be measured by the distances $1 - \bar{B}_2/b_2$ and $1 - \bar{B}_3/b_3$. In this sense, Figure 1 shows that, as expected, the degree of bidispersity grows monotonically as the small-to-big size ratio decreases at a given mole fraction. It also increases as the concentration of the big spheres decreases at a given size ratio, except if the latter ratio is close enough to unity.

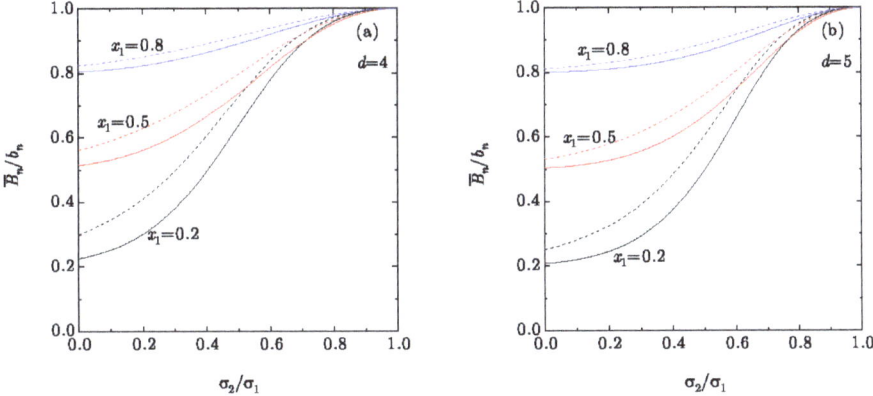

Figure 1. Plot of the ratios $\bar{B}_2/b_2$ (dashed lines) and $\bar{B}_3/b_3$ (solid lines) vs. the size ratio σ_2/σ_1 for binary mixtures with mole fractions $x_1 = 0.2, 0.5,$ and 0.8. Panel (**a**) corresponds to $d=4$, while panel (**b**) corresponds to $d=5$.

To assess the quality of the approximate coefficients (10) and (15), we plot in Figure 2 the ratios B_3^{e1}/B_3 and B_3^{e2}/B_3 as functions of the size ratio σ_2/σ_1 for the same three representative binary compositions as in Figure 1. As we can observe, both the e1 and e2 approximations predict values for the third virial coefficient in overall good agreement with the exact values, especially as the concentration of the big spheres increases. The e1 approximation overestimates B_3 and generally performs worse than the e2 approximation, which tends to overestimate (underestimate) B_3 if the concentration of the big spheres is sufficiently small (large). Additionally, the agreement is better in the four-dimensional case than for five-dimensional hyperspheres. The latter point is relevant because, as said before, the exact expressions of B_3 for $d=4$ are relatively involved [see Equations (18) in the binary case], whereas B_3^{e1} and B_3^{e2} are just simple combinations of moments [see Equations (10a) and (15a)].

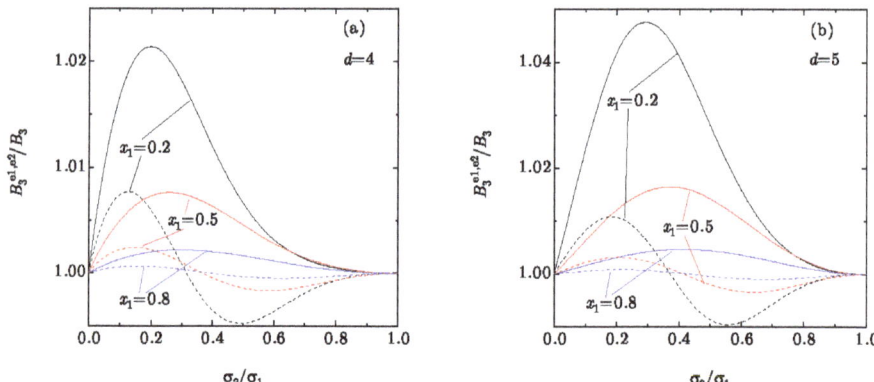

Figure 2. Plot of the ratios B_3^{e1}/B_3 (solid lines) and B_3^{e2}/B_3 (dashed lines) vs. the size ratio σ_2/σ_1 for binary mixtures with mole fractions $x_1 = 0.2$, 0.5, and 0.8. Panel (**a**) corresponds to $d = 4$, while panel (**b**) corresponds to $d = 5$.

The structure of Equation (12) suggests the introduction of a *modified* version (henceforth labeled as "ē1") of the e1 EOS by replacing the approximate third virial coefficient $\bar{B}_3^{\text{e1}}$ by the exact one. More specifically,

$$Z_{\text{ē1}}(\eta) = Z_{\text{e1}}(\eta) + \frac{\bar{B}_3 - \bar{B}_3^{\text{e1}}}{b_3 - b_2}\left[Z_{\text{s}}(\eta) - 1 - b_2\frac{\eta}{1-\eta}\right]. \tag{19}$$

Analogously, we introduce the modified version ("ē2") of the e2 approximation as

$$Z_{\text{ē2}}(\eta) = Z_{\text{e2}}(\eta) + \frac{\bar{B}_3 - \bar{B}_3^{\text{e2}}}{b_3 - b_2}\left[Z_{\text{s}}(\eta) - 1 - b_2\frac{\eta}{1-\eta}\right]. \tag{20}$$

By construction, both $Z_{\text{ē1}}(\eta)$ and $Z_{\text{ē2}}(\eta)$ are consistent with the exact second and third virial coefficients. Moreover, $Z_{\text{ē1}}(\eta) = Z_{\text{ē2}}(\eta)$ for $d = 2$, while $Z_{\text{ē1}}(\eta) = Z_{\text{e1}}(\eta)$ and $Z_{\text{ē2}}(\eta) = Z_{\text{e2}}(\eta)$ for $d = 3$.

2.4. The sp Approximation

Additionally, in previous work [63,64,93], we have adopted an approach to relate the EOS of the polydisperse mixture of d-dimensional hard spheres to the one of the monocomponent fluid which differs from the e1 and e2 approaches in that it does not make use of Equation (1). This involves expressing the excess free energy per particle (a^{ex}) of a polydisperse mixture of packing fraction η in terms of the one of the corresponding monocomponent fluid (a_{s}^{ex}) of an effective packing fraction η_{eff} as

$$\frac{a^{\text{ex}}(\eta)}{k_B T} + \ln(1-\eta) = \frac{\alpha}{\lambda}\left[\frac{a_{\text{s}}^{\text{ex}}(\eta_{\text{eff}})}{k_B T} + \ln(1-\eta_{\text{eff}})\right]. \tag{21}$$

In Equation (21), η_{eff} and η are related through

$$\frac{\eta_{\text{eff}}}{1-\eta_{\text{eff}}} = \frac{1}{\lambda}\frac{\eta}{1-\eta}, \quad \eta_{\text{eff}} = \left[1+\lambda\left(\eta^{-1}-1\right)\right]^{-1}, \tag{22}$$

while the parameters λ and α are determined by imposing consistency with the (exact) second and third virial coefficients of the mixture, Equations (9) and (16). More specifically [63,64],

$$\lambda = \frac{\bar{B}_2 - 1}{b_2 - 1}\frac{b_3 - 2b_2 + 1}{\bar{B}_3 - 2\bar{B}_2 + 1}, \quad \alpha = \lambda^2\frac{\bar{B}_2 - 1}{b_2 - 1}. \tag{23}$$

Note that the ratio $\eta/(1-\eta)$ represents a rescaled packing fraction, i.e., the ratio between the volume occupied by the spheres and the remaining void volume. Thus, according to Equation (22), the effective monocomponent fluid associated with a given mixture has a rescaled packing fraction $\eta_{\text{eff}}/(1-\eta_{\text{eff}})$ that is λ times smaller than that of the mixture. Moreover, in the case of three-dimensional hard-sphere mixtures, Equations (21)–(23) can be derived in the context of consistent fundamental-measure theories [63,64,97,98].

Taking into account the thermodynamic relation

$$Z(\eta) = 1 + \eta \frac{\partial a^{\text{ex}}(\eta)/k_B T}{\partial \eta}, \tag{24}$$

the mapping between the compressibility factor of the d-dimensional monocomponent system (Z_s) and the approximate one of the polydisperse mixture that is then obtained from Equation (21) may be expressed as

$$\eta Z_{\text{sp}}(\eta) - \frac{\eta}{1-\eta} = \alpha \left[\eta_{\text{eff}} Z_s(\eta_{\text{eff}}) - \frac{\eta_{\text{eff}}}{1-\eta_{\text{eff}}} \right], \tag{25}$$

where a label "sp", motivated by the nomenclature already introduced in connection with the "surplus" pressure $\eta Z(\eta) - \eta/(1-\eta)$ [63], has been added to distinguish this compressibility factor from the previous approximations.

Equation (25) shares with Equations (19) and (20) the consistency with the exact second and third virial coefficients. On the other hand, while $Z_{\tilde{e}1}(\eta)$ and $Z_{\tilde{e}2}(\eta)$ are related to the monocomponent compressibility factor $Z_s(\eta)$ evaluated at the same packing fraction η as that of the mixture, $Z_{\text{sp}}(\eta)$ is related to $Z_s(\eta_{\text{eff}})$ evaluated at a different (effective) packing fraction η_{eff}.

Figure 3 shows that $\lambda > 1$, while $\alpha < 1$, except if the mole fraction of the big spheres is large enough (not shown). According to Equations (22) and (25), this implies that (i) $\eta_{\text{eff}} < \eta$ and (ii) the surplus pressure of the mixture at a packing fraction η is generally smaller than that of the monocomponent fluid at the equivalent packing fraction η_{eff}. It is also worthwhile noting that, in contrast to what happens with $\bar{B}_2$ and $\bar{B}_3$ (see Figure 1), λ has a nonmonotonic dependence on the size ratio and α also exhibits a nonmonotonic behavior if x_1 is small enough.

While we have proved the sp approach to be successful for both hard-disk ($d = 2$) [64] and hard-sphere ($d = 3$) [93] mixtures, one of our goals is to test it for $d = 4$ and $d = 5$ as well.

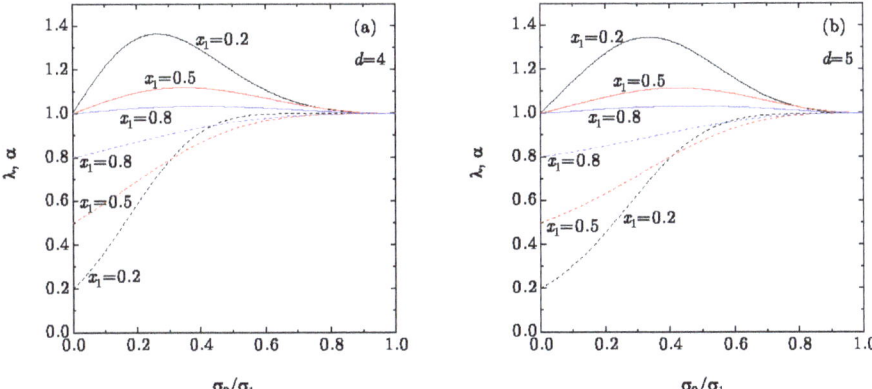

Figure 3. Plot of the coefficients λ (solid lines) and α (dashed lines) [see Equation (23)] vs. the size ratio σ_2/σ_1 for binary mixtures with mole fractions $x_1 = 0.2, 0.5$, and 0.8. Panel (**a**) corresponds to $d = 4$, while panel (**b**) corresponds to $d = 5$.

3. Comparison with Computer Simulation Results

In order to obtain explicit numerical results for the different approximations to the EOS of four- and five-dimensional hard-sphere mixtures, we require an expression for $Z_s(\eta)$. While other choices are available, we considered here the empirical proposal that works for both dimensionalities by Luban and Michels (LM) [25], which reads

$$Z_s(\eta) = 1 + b_2 \eta \frac{1 + [b_3/b_2 - \zeta(\eta) b_4/b_3] \eta}{1 - \zeta(\eta)(b_4/b_3)\eta + [\zeta(\eta) - 1](b_4/b_2)\eta^2}, \qquad (26)$$

where $\zeta(\eta) = \zeta_0 + \zeta_1 \eta/\eta_{cp}$, η_{cp} being the crystalline close-packing value. The values of b_2, b_3, b_4, ζ_0, ζ_1, and η_{cp} are given in Table 1.

Table 1. Values of b_2–b_4, ζ_0, ζ_1, and η_{cp} for $d = 4$ and 5.

	$d = 4$	$d = 5$
b_2	8	16
b_3	$2^6 \left(\frac{4}{3} - \frac{3\sqrt{3}}{2\pi}\right) \simeq 32.406$	106
b_4	$2^9 \left(2 - \frac{27\sqrt{3}}{4\pi} + \frac{832}{45\pi^2}\right) \simeq 77.7452$	$\frac{25\,315\,393}{8008} + \frac{3\,888\,425\sqrt{2}}{4\,004\pi} - \frac{67\,183\,425\arccos(1/3)}{8008\pi} \simeq 311.183$
ζ_0	1.2973(59)	1.074(16)
ζ_1	$-0.062(13)$	0.163(45)
η_{cp}	$\frac{\pi^2}{16} \simeq 0.617$	$\frac{\pi^2\sqrt{2}}{30} \simeq 0.465$

In Table 2 we list the systems whose compressibility factor has been obtained from simulation, either using MD [36] or MC [57,59] methods. The values of the corresponding coefficients $\tilde{B}_2$ [see Equation (9)], $\tilde{B}_3$ [see Equations (16)–(18)], λ, and α [see Equation (23)] are also included. We assigned a three-character label to each system, where the first (capital) letter denotes the size ratio (A–F for $\sigma_2/\sigma_1 = \frac{1}{4}, \frac{1}{3}, \frac{2}{5}, \frac{1}{2}, \frac{3}{5}$, and $\frac{3}{4}$, respectively), the second (lower-case) letter denotes the mole fraction (a, b, and c for $x_1 = 0.25, 0.50$, and 0.75, respectively), and the digit (4 or 5) denotes the dimensionality.

Table 2. Binary mixtures of four- and five-dimensional hard spheres studied through simulations (Monte Carlo—MC or molecular dynamics—MD) and the values of their coefficients $\bar{B}_2$ [see Equation (9)], $\bar{B}_3$ [see Equations (16)–(18)], λ, and α [see Equation (23)].

d	Label	σ_2/σ_1	x_1	Simulation Method	$\bar{B}_2$	$\bar{B}_3$	λ	α
4	Aa4	1/4	0.25	MD [1]	3.85618	12.2253	1.28824	0.677138
	Ab4	1/4	0.50	MD [1]	5.21595	18.8828	1.10923	0.741033
	Ac4	1/4	0.75	MD [1]	6.60436	25.6326	1.03810	0.862800
	Ba4	1/3	0.25	MD [1]	4.42857	14.4931	1.28470	0.808392
	Bb4	1/3	0.50	MD [1]	5.56098	20.2530	1.11943	0.816497
	Bc4	1/3	0.75	MD [1]	6.77049	26.2935	1.04334	0.897356
	Cb4	2/5	0.50	MC [2]	5.87285	21.5939	1.11692	0.868418
	Da4	1/2	0.25	MD [1]	5.82895	20.8444	1.17876	0.958523
	Db4	1/2	0.50	MD [1] and MC [2]	6.38235	23.9444	1.09883	0.928396
	Dc4	1/2	0.75	MD [1]	7.15816	28.0333	1.04047	0.952376
	Eb4	3/5	0.50	MC [2]	6.90085	26.5045	1.07078	0.966532
	Fa4	3/4	0.25	MD [1]	7.55661	29.9061	1.03231	0.998173
	Fb4	3/4	0.50	MD [1]	7.56231	29.9832	1.02894	0.992515
	Fc4	3/4	0.75	MD [1]	7.73940	30.9790	1.01561	0.993060
5	Aa5	1/4	0.25	MD [1]	6.30550	32.9426	1.24358	0.546995
	Ab5	1/4	0.50	MD [1]	9.52439	57.2455	1.08739	0.671954
	Ac5	1/4	0.75	MD [1]	12.7601	81.6145	1.02988	0.831562
	Ba5	1/3	0.25	MD [1]	7.21951	37.7995	1.27656	0.675687
	Bb5	1/3	0.50	MD [1]	10.0984	60.3097	1.10651	0.742645
	Bc5	1/3	0.75	MD [1]	13.0411	83.1175	1.03739	0.863898
	Cb5	2/5	0.50	MC [3,4]	10.6565	63.6666	1.11369	0.798464
	Da5	1/2	0.25	MD [1]	9.89286	55.1378	1.22316	0.886983
	Db5	1/2	0.50	MD [1] and MC [3,5]	11.6818	70.5615	1.10812	0.874437
	Dc5	1/2	0.75	MD [1]	13.7964	88.0120	1.04172	0.925768
	Fa5	3/4	0.25	MD [1]	14.5176	92.4875	1.04866	0.990981
	Fb5	3/4	0.50	MD [1]	14.6327	93.8346	1.03957	0.982162
	Fc5	3/4	0.75	MD [1]	15.2162	99.1168	1.02005	0.986104

[1] Ref. [36], [2] Ref. [57], [3] Ref. [59], [4] $x_1 = \frac{971}{1944} = 0.499486$, [5] $x_1 = \frac{973}{1944} = 0.500514$.

If, as before, the degree of bidispersity is measured by $1 - \bar{B}_2/b_2$ and $1 - \bar{B}_3/b_3$, we can observe the following ordering of decreasing bidispersity in the four-dimensional systems: Aa, Ba, Ab, Bb, Da, Cb, Db, Ac, Bc, Eb, Dc, Fa, Fb, and Fc. The same ordering applies in the case of the five-dimensional systems, except that, apart from the absence of the system Eb, the sequence {Ab, Bb, Da} is replaced by either {Ab, Da, Bb} or by {Da, Ab, Bb} if either $1 - \bar{B}_2/b_2$ or $1 - \bar{B}_3/b_3$ are used, respectively.

It should be stressed that the proposals implied by Equations (4), (13), (19), (20), and (25) may be interpreted in two directions. On the one hand, if Z_s is known as a function of the packing fraction, then one can readily compute the compressibility factor of the mixture for any packing fraction and composition [η_{eff} and η being related through Equation (22) in the case of Z_{sp}]; this is the standard view. On the other hand, if simulation data for the EOS of the mixture are available for different densities, size ratios, and mole fractions, Equations (4), (13), (19), (20), and (25) can be used to *infer* the compressibility factor of the monocomponent fluid. This is particularly important in the high-density region, where obtaining data from simulation may be accessible in the case of mixtures but either difficult or not feasible in the case of the monocomponent fluid, as happens in the metastable fluid branch [64,93].

In principle, simulation data for different mixtures would yield different inferred functions $Z_s(\eta)$. Thus, without having to use an externally imposed monocomponent EOS, the degree of collapse of the mapping from mixture compressibility factors onto a *common* function $Z_s(\eta)$ is an efficient way of assessing the performance of Equations (4), (13), (19), (20), and (25). As shown in Figure 4, the usefulness of those mappings is confirmed by the nice collapse obtained for all the points corresponding to the mixtures described in Table 2. The inferred data associated with $Z_{\bar{e}2}$ are almost identical to those associated with Z_{e2} and thus they are omitted in Figure 4. Figure 4 also shows that the inferred curves are very close to the LM (monocomponent) EOS, Equation (26), what validates its choice as an accurate function $Z_s(\eta)$ in what follows. Notwithstanding this, one can observe in the high-density regime that the values inferred from simulation data via Z_{e1} and $Z_{\bar{e}1}$ tend to underestimate the LM curve for both $d=4$ and $d=5$, while the values inferred via Z_{e2} tend to overestimate it for $d=5$. Overall, one can say that the best agreement with the LM EOS is obtained by using Z_{e2} and Z_{sp} for $d=4$ and $d=5$, respectively.

Now we turn to a more a direct comparison between the simulation data and the approximate EOS for mixtures. As expected from the indirect representation of Figure 4, we observed a very good agreement (not shown) between the simulation data for the systems displayed in Table 2 and the theoretical predictions obtained from Equations (4), (13), (19), (20), and (25), supplemented by Equation (26).

In order to perform a more stringent assessment of the five theoretical EOS, we chose $Z_{e1}(\eta)$ as a *reference* theory and focused on the percentage deviation $100[Z(\eta)/Z_{e1}(\eta) - 1]$ from it. The results are displayed in Figures 5 and 6 for $d=4$ and Figures 7 and 8 for $d=5$. Those figures reinforce the view that all our theoretical proposals are rather accurate: the errors in Z_{e1} were typically smaller than 1% and they are even smaller in the other approximate EOS. Note that we have not put error bars in the MD data since they were unfortunately not reported in Reference [36]. We must also mention that the MD data were generally more scattered than the MC ones. Moreover, certain (small) discrepancies between MC and MD points can be observed in Figure 6c, MC data generally lying below MD data. The same feature is also present (although somewhat less apparent) in Figure 8c. This may be due to larger finite-size effects in the MD simulations than in the MC simulations: the MD simulations used 648 hyperspheres for $d=4$ and 512 or 1024 hyperspheres for $d=5$, while the MC simulations used 10,000 hyperspheres for $d=4$ and 3888 or 7776 for $d=5$. In any case, since the MC data were statistically precise, the discrepancy might be eliminated by the inclusion of the (unknown) error bars in the MD results. It is also worth pointing out that the representation of Figures 5–8 is much more demanding than a conventional representation of Z vs. η for each mixture or even the representation of Figure 4.

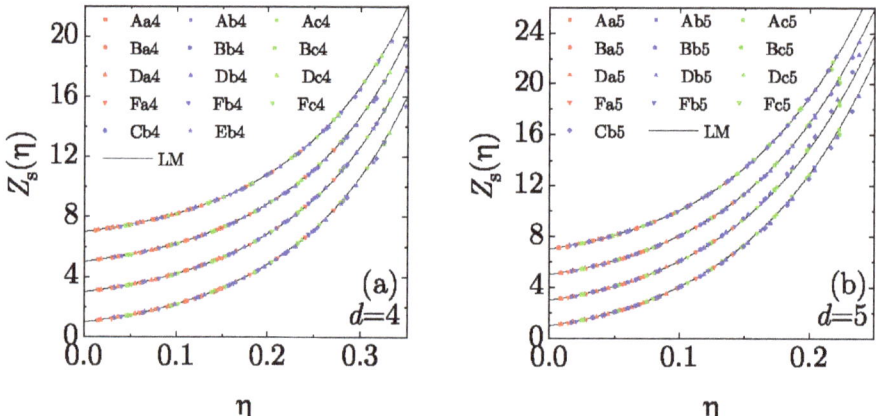

Figure 4. Plot of the monocomponent compressibility factor $Z_s(\eta)$, as inferred from simulation data for the mixtures described in Table 2, according to the theories (from bottom to top) e1, e2, ē1, and sp (the three latter have been shifted vertically for better clarity). The solid lines represent the Luban and Michels (LM) equation of state (EOS), Equation (26). Panel (**a**) corresponds to $d=4$, while panel (**b**) corresponds to $d=5$.

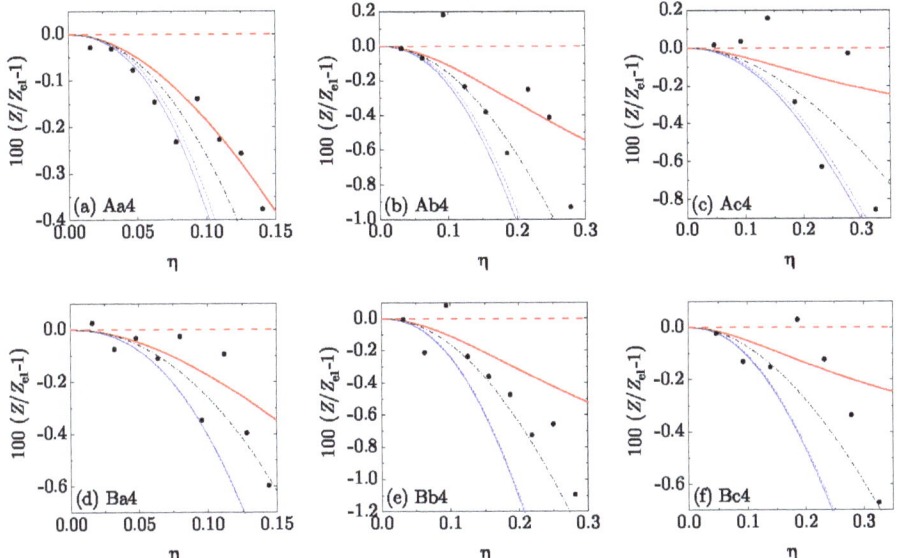

Figure 5. Plot of the relative deviations $100[Z(\eta)/Z_{e1}(\eta) - 1]$ from the theoretical EOS $Z_{e1}(\eta)$ for the four-dimensional mixtures Aa4–Bc4 (see Table 2). Thick (red) dashed lines: e1; thick (red) solid lines: ē1; thin (blue) dashed lines: e2; thin (blue) solid lines: ē2; dash-dotted (black) lines: sp; filled (black) circles: MD.

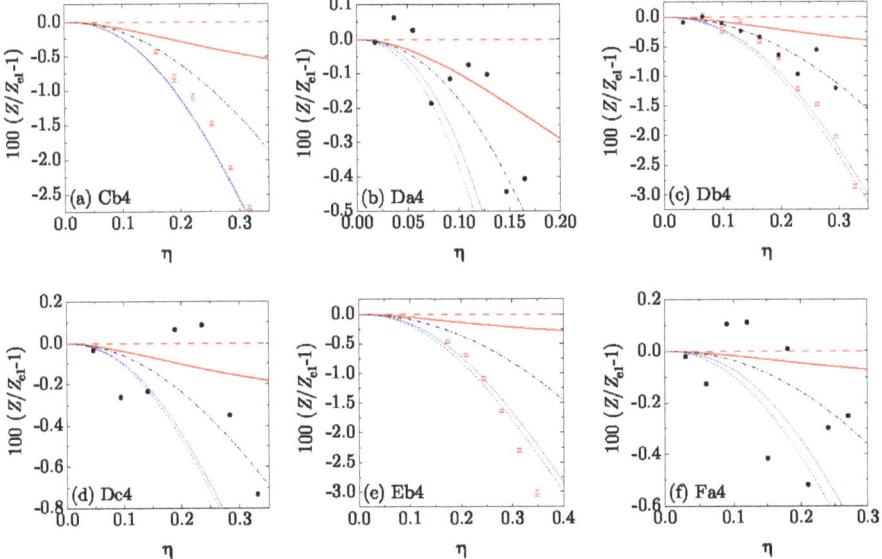

Figure 6. Cont.

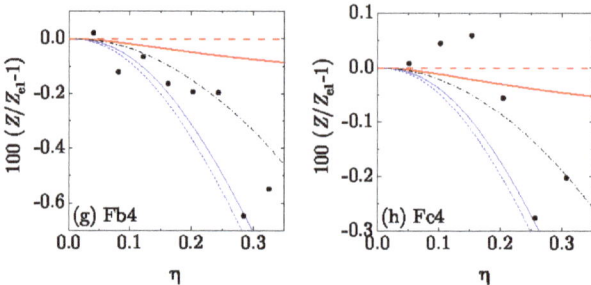

Figure 6. Plot of the relative deviations $100[Z(\eta)/Z_{e1}(\eta) - 1]$ from the theoretical EOS $Z_{e1}(\eta)$ for the four-dimensional mixtures Cb4–Fc4 (see Table 2). Thick (red) dashed lines: e1; thick (red) solid lines: ē1; thin (blue) dashed lines: e2; thin (blue) solid lines: ē2; dash-dotted (black) lines: sp; filled (black) circles: MD; open (red) triangles with error bars in panels (**a**), (**c**), and (**e**): MC.

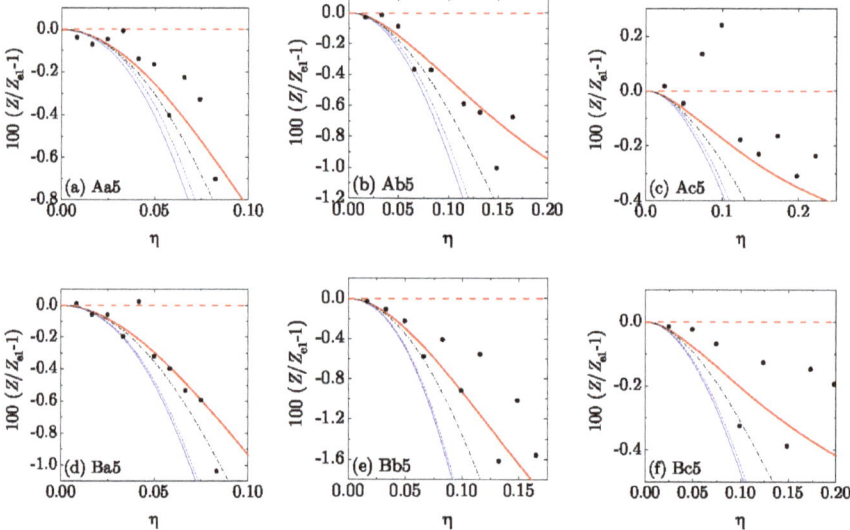

Figure 7. Plot of the relative deviations $100[Z(\eta)/Z_{e1}(\eta) - 1]$ from the theoretical EOS $Z_{e1}(\eta)$ for the five-dimensional mixtures Aa5–Bc5 (see Table 2). Thick (red) dashed lines: e1; thick (red) solid lines: ē1; thin (blue) dashed lines: e2; thin (blue) solid lines: ē2; dash-dotted (black) lines: sp; filled (black) circles: MD.

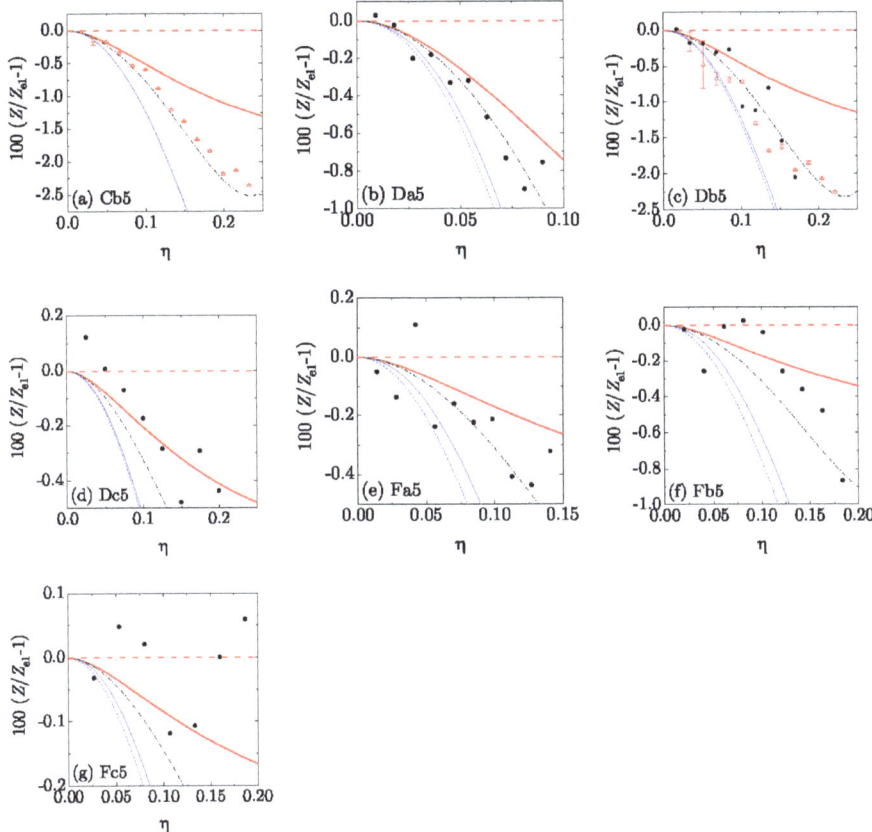

Figure 8. Plot of the relative deviations $100[Z(\eta)/Z_{e1}(\eta) - 1]$ from the theoretical EOS $Z_{e1}(\eta)$ for the five-dimensional mixtures Cb5–Fc5 (see Table 2). Thick (red) dashed lines: e1; thick (red) solid lines: ē1; thin (blue) dashed lines: e2; thin (blue) solid lines: ē2; dash-dotted (black) lines: sp; filled (black) circles: MD; open (red) triangles with error bars in panels (**a**) and (**c**): MC.

4. Discussion and Concluding Remarks

In this paper we have carried out a thorough comparison between our theoretical proposals for the EOS of a multicomponent d-dimensional mixture of hard hyperspheres and the available simulation results for binary mixtures of both four- and five-dimensional hard hyperspheres. It should be stressed that in this comparison we have restricted ourselves to the liquid branch. Let us now summarize the outcome of the different theories for the compressibility factor.

First, we note that $Z_{\bar{e}2}(\eta) \approx Z_{e2}(\eta) < Z_{sp}(\eta) < Z_{\bar{e}1}(\eta) < Z_{e1}(\eta)$. The fact that $Z_{\bar{e}2}(\eta) \approx Z_{e2}(\eta)$ is a consequence of the small deviations of B_3^{e2} from the exact third virial coefficient (see Figure 2). Thus, there does not seem to be any practical advantage in choosing $Z_{\bar{e}2}$ instead of Z_{e2}, especially if $d = 4$ [where the exact B_3 has a rather involved expression, see Equations (18)]. If one restricts oneself to the comparison between those approximate EOS that do not yield the exact B_3, namely Z_{e1} and Z_{e2}, we find that Z_{e2} performs generally better. On the other hand, if approximations requiring the exact B_3 as input are considered, namely $Z_{\bar{e}1}$, $Z_{\bar{e}2}$, and Z_{sp}, the conclusion is that Z_{sp} generally outperforms the other two.

The comparison with the simulation data confirms that the good agreement between the results of $Z_{e1}(\eta)$ that had been found earlier in connection with both MD [36] and MC [57,59] simulation data are even improved by the other approximate theories. In fact, in both the four- and five-dimensional cases, the best agreement with the MD results is generally obtained from Z_{e1} and Z_{sp}. On the other hand, for the four-dimensional case, the best agreement with the MC results corresponds to $Z_{e2} \approx Z_{e2}$, while that for the five-dimensional case corresponds to Z_{sp}.

Finally, it must be pointed out that it seems that overall Z_{sp} exhibits the best global behavior. However, more accurate simulation data would be needed to confirm this conclusion. It should also be stressed that the performance of the analyzed approximate EOS for fluid mixtures might be affected by the reliability of the (monocomponent) LM EOS. In any event, one may reasonably argue that the mapping between the compressibility factor of the mixture and the one of the monocomponent system with an effective packing fraction [see Equations (22) and (25)] that had already been tested in two- [64] and three-dimensional [93] mixtures is confirmed as an excellent approach also for higher dimensions.

Author Contributions: A.S. proposed the idea and the three authors performed the calculations. The three authors also participated in the analysis and discussion of the results and worked on the revision and writing of the final manuscript. All authors have read and agreed to the published version of the manuscript.

Funding: A.S. and S.B.Y. acknowledge financial support from the Spanish Agencia Estatal de Investigación through Grant No. FIS2016-76359-P and the Junta de Extremadura (Spain) through Grant No. GR18079, both partially financed by Fondo Europeo de Desarrollo Regional funds.

Conflicts of Interest: The authors declare no conflict of interest.

Abbreviations

The following abbreviations are used in this manuscript:

EOS   Equation of state
LM   Luban–Michels
MC   Monte Carlo
MD   Molecular dynamics

References

1. Frisch, H.L.; Rivier, N.; Wyler, D. Classical Hard-Sphere Fluid in Infinitely Many Dimensions. *Phys. Rev. Lett.* **1985**, *54*, 2061–2063. [CrossRef] [PubMed]
2. Luban, M. Comment on "Classical Hard-Sphere Fluid in Infinitely Many Dimensions". *Phys. Rev. Lett.* **1986**, *56*, 2330–2330. [CrossRef] [PubMed]
3. Frisch, H.L.; Rivier, N.; Wyler, D. Frisch, Rivier, and Wyler Respond. *Phys. Rev. Lett.* **1986**, *56*, 2331–2331. [CrossRef] [PubMed]
4. Klein, W.; Frisch, H.L. Instability in the infinite dimensional hard-sphere fluid. *J. Chem. Phys.* **1986**, *84*, 968–970. [CrossRef]
5. Wyler, D.; Rivier, N.; Frisch, H.L. Hard-sphere fluid in infinite dimensions. *Phys. Rev. A* **1987**, *36*, 2422–2431. [CrossRef] [PubMed]
6. Bagchi, B.; Rice, S.A. On the stability of the infinite dimensional fluid of hard hyperspheres: A statistical mechanical estimate of the density of closest packing of simple hypercubic lattices in spaces of large dimensionality. *J. Chem. Phys.* **1988**, *88*, 1177–1184. [CrossRef]
7. Elskens, Y.; Frisch, H.L. Kinetic theory of hard spheres in infinite dimensions. *Phys. Rev. A* **1988**, *37*, 4351–4353. [CrossRef]
8. Carmesin, H.O.; Frisch, H.; Percus, J. Binary nonadditive hard-sphere mixtures at high dimension. *J. Stat. Phys.* **1991**, *63*, 791–795. [CrossRef]
9. Frisch, H.L.; Percus, J.K. High dimensionality as an organizing device for classical fluids. *Phys. Rev. E* **1999**, *60*, 2942–2948. [CrossRef] [PubMed]
10. Parisi, G.; Slanina, F. Toy model for the mean-field theory of hard-sphere liquids. *Phys. Rev. E* **2000**, *62*, 6554–6559. [CrossRef]
11. Yukhimets, A.; Frisch, H.L.; Percus, J.K. Molecular Fluids at High Dimensionality. *J. Stat. Phys.* **2000**, *100*, 135–151. [CrossRef]

12. Charbonneau, P.; Kurchan, J.; Parisi, G.; Urbani, P.; Zamponi, F. Glass and Jamming Transitions: From Exact Results to Finite-Dimensional Descriptions. *Annu. Rev. Cond. Matter Phys.* **2017**, *8*, 265–288. [CrossRef]
13. Santos, A.; López de Haro, M. Demixing can occur in binary hard-sphere mixtures with negative non-additivity. *Phys. Rev. E* **2005**, *72*, 010501(R). [CrossRef]
14. Freasier, C.; Isbister, D.J. A remark on the Percus–Yevick approximation in high dimensions. Hard core systems. *Mol. Phys.* **1981**, *42*, 927–936. [CrossRef]
15. Leutheusser, E. Exact solution of the Percus–Yevick equation for a hard-core fluid in odd dimensions. *Physica A* **1984**, *127*, 667–676. [CrossRef]
16. Michels, J.P.J.; Trappeniers, N.J. Dynamical computer simulations on hard hyperspheres in four- and five-dimensional space. *Phys. Lett. A* **1984**, *104*, 425–429. [CrossRef]
17. Baus, M.; Colot, J.L. Theoretical structure factors for hard-core fluids. *J. Phys. C* **1986**, *19*, L643–L648. [CrossRef]
18. Baus, M.; Colot, J.L. Thermodynamics and structure of a fluid of hard rods, disks, spheres, or hyperspheres from rescaled virial expansions. *Phys. Rev. A* **1987**, *36*, 3912–3925. [CrossRef]
19. Rosenfeld, Y. Distribution function of two cavities and Percus–Yevick direct correlation functions for a hard sphere fluid in D dimensions: Overlap volume function representation. *J. Chem. Phys.* **1987**, *87*, 4865–4869. [CrossRef]
20. Rosenfeld, Y. Scaled field particle theory of the structure and thermodynamics of isotropic hard particle fluids. *J. Chem. Phys.* **1988**, *89*, 4272–4287. [CrossRef]
21. Amorós, J.; Solana, J.R.; Villar, E. Equations of state for four- and five-dimensional hard hypersphere fluids. *Phys. Chem. Liq.* **1989**, *19*, 119–124. [CrossRef]
22. Song, Y.; Mason, E.A.; Stratt, R.M. Why does the Carnahan-Starling equation work so well? *J. Phys. Chem.* **1989**, *93*, 6916–6919. [CrossRef]
23. Song, Y.; Mason, E.A. Equation of state for fluids of spherical particles in d dimensions. *J. Chem. Phys.* **1990**, *93*, 686–688. [CrossRef]
24. González, D.J.; González, L.E.; Silbert, M. Thermodynamics of a fluid of hard D-dimensional spheres: Percus-Yevick and Carnahan-Starling-like results for $D = 4$ and 5. *Phys. Chem. Liq.* **1990**, *22*, 95–102. [CrossRef]
25. Luban, M.; Michels, J.P.J. Equation of state of hard D-dimensional hyperspheres. *Phys. Rev. A* **1990**, *41*, 6796–6804. [CrossRef]
26. Maeso, M.J.; Solana, J.R.; Amorós, J.; Villar, E. Equations of state for D-dimensional hard sphere fluids. *Mater. Chem. Phys.* **1991**, *30*, 39–42. [CrossRef]
27. González, D.J.; González, L.E.; Silbert, M. Structure and thermodynamics of hard D-dimensional spheres: overlap volume function approach. *Mol. Phys.* **1991**, *74*, 613–627. [CrossRef]
28. González, L.E.; González, D.J.; Silbert, M. Structure and thermodynamics of mixtures of hard D-dimensional spheres: Overlap volume function approach. *J. Chem. Phys.* **1992**, *97*, 5132–5141. [CrossRef]
29. Velasco, E.; Mederos, L.; Navascués, G. Analytical approach to the thermodynamics and density distribution of crystalline phases of hard spheres spheres. *Mol. Phys.* **1999**, *97*, 1273–1277. [CrossRef]
30. Bishop, M.; Masters, A.; Clarke, J.H.R. Equation of state of hard and Weeks–Chandle–Anderson hyperspheres in four and five dimensions. *J. Chem. Phys.* **1999**, *110*, 11449–11453. [CrossRef]
31. Finken, R.; Schmidt, M.; Löwen, H. Freezing transition of hard hyperspheres. *Phys. Rev. E* **2001**, *65*, 016108. [CrossRef]
32. Santos, A.; Yuste, S.B.; López de Haro, M. Equation of state of a multicomponent d-dimensional hard-sphere fluid. *Mol. Phys.* **1999**, *96*, 1–5. [CrossRef]
33. Mon, K.K.; Percus, J.K. Virial expansion and liquid-vapor critical points of high dimension classical fluids. *J. Chem. Phys.* **1999**, *110*, 2734–2735. [CrossRef]
34. Santos, A. An equation of state à La Carnahan-Starling A Five-Dimens. Fluid Hard Hyperspheres. *J. Chem. Phys.* **2000**, *112*, 10680–10681. [CrossRef]
35. Yuste, S.B.; Santos, A.; López de Haro, M. Demixing in binary mixtures of hard hyperspheres. *Europhys. Lett.* **2000**, *52*, 158–164. [CrossRef]
36. González-Melchor, M.; Alejandre, J.; López de Haro, M. Equation of state and structure of binary mixtures of hard d-dimensional hyperspheres. *J. Chem. Phys.* **2001**, *114*, 4905–4911. [CrossRef]

37. Santos, A.; Yuste, S.B.; López de Haro, M. Contact values of the radial distribution functions of additive hard-sphere mixtures in d dimensions: A new proposal. *J. Chem. Phys.* **2002**, *117*, 5785–5793. [CrossRef]
38. Robles, M.; López de Haro, M.; Santos, A. Equation of state of a seven-dimensional hard-sphere fluid. Percus–Yevick theory and molecular-dynamics simulations. *J. Chem. Phys.* **2004**, *120*, 9113–9122. [CrossRef]
39. Santos, A.; López de Haro, M.; Yuste, S.B. Equation of state of nonadditive d-dimensional hard-sphere mixtures. *J. Chem. Phys.* **2005**, *122*, 024514. [CrossRef]
40. Bishop, M.; Whitlock, P.A.; Klein, D. The structure of hyperspherical fluids in various dimensions. *J. Chem. Phys.* **2005**, *122*, 074508. [CrossRef]
41. Bishop, M.; Whitlock, P.A. The equation of state of hard hyperspheres in four and five dimensions. *J. Chem. Phys.* **2005**, *123*, 014507. [CrossRef]
42. Lue, L.; Bishop, M. Molecular dynamics study of the thermodynamics and transport coefficients of hard hyperspheres in six and seven dimensions. *Phys. Rev. E* **2006**, *74*, 021201. [CrossRef] [PubMed]
43. López de Haro, M.; Yuste, S.B.; Santos, A. Test of a universality ansatz for the contact values of the radial distribution functions of hard-sphere mixtures near a hard wall. *Mol. Phys.* **2006**, *104*, 3461–3467. [CrossRef]
44. Bishop, M.; Whitlock, P.A. Monte Carlo Simulation of Hard Hyperspheres in Six, Seven and Eight Dimensions for Low to Moderate Densities. *J. Stat. Phys.* **2007**, *126*, 299–314. [CrossRef]
45. Robles, M.; López de Haro, M.; Santos, A. Percus–Yevick theory for the structural properties of the seven-dimensional hard-sphere fluid. *J. Chem. Phys.* **2007**, *126*, 016101. [CrossRef] [PubMed]
46. Whitlock, P.A.; Bishop, M.; Tiglias, J.L. Structure factor for hard hyperspheres in higher dimensions. *J. Chem. Phys.* **2007**, *126*, 224505. [CrossRef] [PubMed]
47. Rohrmann, R.D.; Santos, A. Structure of hard-hypersphere fluids in odd dimensions. *Phys. Rev. E* **2007**, *76*, 051202. [CrossRef] [PubMed]
48. López de Haro, M.; Yuste, S.B.; Santos, A. Alternative Approaches to the Equilibrium Properties of Hard-Sphere Liquids. In *Theory and Simulation of Hard-Sphere Fluids and Related Systems*; Mulero, A., Ed.; Lecture Notes in Physics; Springer: Berlin, Germany, 2008; Volume 753, pp. 183–245.
49. Bishop, M.; Clisby, N.; Whitlock, P.A. The equation of state of hard hyperspheres in nine dimensions for low to moderate densities. *J. Chem. Phys.* **2008**, *128*, 034506. [CrossRef]
50. Adda-Bedia, M.; Katzav, E.; Vella, D. Solution of the Percus–Yevick equation for hard hyperspheres in even dimensions. *J. Chem. Phys.* **2008**, *129*, 144506. [CrossRef]
51. Rohrmann, R.D.; Robles, M.; López de Haro, M.; Santos, A. Virial series for fluids of hard hyperspheres in odd dimensions. *J. Chem. Phys.* **2008**, *129*, 014510. [CrossRef]
52. van Meel, J.A.; Charbonneau, B.; Fortini, A.; Charbonneau, P. Hard-sphere crystallization gets rarer with increasing dimension. *Phys. Rev. E* **2009**, *80*, 061110. [CrossRef] [PubMed]
53. Lue, L.; Bishop, M.; Whitlock, P.A. The fluid to solid phase transition of hard hyperspheres in four and five dimensions. *J. Chem. Phys.* **2010**, *132*, 104509. [CrossRef] [PubMed]
54. Rohrmann, R.D.; Santos, A. Multicomponent fluids of hard hyperspheres in odd dimensions. *Phys. Rev. E* **2011**, *83*, 011201. [CrossRef] [PubMed]
55. Leithall, G.; Schmidt, M. Density functional for hard hyperspheres from a tensorial-diagrammatic series. *Phys. Rev. E* **2011**, *83*, 021201. [CrossRef] [PubMed]
56. Estrada, C.D.; Robles, M. Fluid–solid transition in hard hypersphere systems. *J. Chem. Phys.* **2011**, *134*, 044115. [CrossRef] [PubMed]
57. Bishop, M.; Whitlock, P.A. Monte Carlo study of four dimensional binary hard hypersphere mixtures. *J. Chem. Phys.* **2012**, *136*, 014506. [CrossRef]
58. Bishop, M.; Whitlock, P.A. Phase transitions in four-dimensional binary hard hypersphere mixtures. *J. Chem. Phys.* **2013**, *138*, 084502. [CrossRef]
59. Bishop, M.; Whitlock, P.A. Five dimensional binary hard hypersphere mixtures: A Monte Carlo study. *J. Chem. Phys.* **2016**, *145*, 154502. [CrossRef]
60. Amorós, J.; Ravi, S. On the application of the Carnahan–Starling method for hard hyperspheres in several dimensions. *Phys. Lett. A* **2013**, *377*, 2089–2092. [CrossRef]
61. Amorós, J. Equations of state for tetra-dimensional hard-sphere fluids. *Phys. Chem. Liq.* **2014**, *52*, 287–290. [CrossRef]
62. Heinen, M.; Horbach, J.; Löwen, H. Liquid pair correlations in four spatial dimensions: Theory versus simulation. *Mol. Phys.* **2015**, *113*, 1164–1169. [CrossRef]

63. Santos, A. *A Concise Course on the Theory of Classical Liquids. Basics and Selected Topics*; Lecture Notes in Physics; Springer: New York, NY, USA, 2016; Volume 923.
64. Santos, A.; Yuste, S.B.; López de Haro, M.; Ogarko, V. Equation of state of polydisperse hard-disk mixtures in the high-density regime. *Phys. Rev. E* **2017**, *93*, 062603. [CrossRef] [PubMed]
65. Akhouri, B.P. Equations of state for hard hypersphere fluids in high dimensional spaces. *Int. J. Chem. Stud.* **2017**, *5*, 39–45. [CrossRef]
66. Ivanizki, D. A generalization of the Carnahan–Starling approach with applications to four- and five-dimensional hard spheres. *Phys. Lett. A* **2018**, *382*, 1745–1751. [CrossRef]
67. Santos, A.; Yuste, S.B.; López de Haro, M. Virial coefficients and equations of state for mixtures of hard discs, hard spheres, and hard hyperspheres. *Mol. Phys.* **2001**, *99*, 1959–1972. [CrossRef]
68. Ree, F.H.; Hoover, W.G. On the Signs of the Hard Sphere Virial Coefficients. *J. Chem. Phys.* **1964**, *40*, 2048–2049. [CrossRef]
69. Luban, M.; Baram, A. Third and fourth virial coefficients of hard hyperspheres of arbitrary dimensionality. *J. Chem. Phys.* **1982**, *76*, 3233–3241. [CrossRef]
70. Joslin, C.G. Third and fourth virial coefficients of hard hyperspheres of arbitrary dimensionality. *J. Chem. Phys.* **1982**, *77*, 2701–2702. [CrossRef]
71. Loeser, J.G.; Zhen, Z.; Kais, S.; Herschbach, D.R. Dimensional interpolation of hard sphere virial coefficients. *J. Chem. Phys.* **1991**, *95*, 4525–4544. [CrossRef]
72. Enciso, E.; Almarza, N.G.; González, M.A.; Bermejo, F.J. The virial coefficients of hard hypersphere binary mixtures. *Mol. Phys.* **2002**, *100*, 1941–1944. [CrossRef]
73. Bishop, M.; Masters, A.; Vlasov, A.Y. Higher virial coefficients of four and five dimensional hard hyperspheres. *J. Chem. Phys.* **2004**, *121*, 6884–6886. [CrossRef] [PubMed]
74. Clisby, N.; McCoy, B.M. Analytic Calculation of B_4 for Hard Spheres in Even Dimensions. *J. Stat. Phys.* **2004**, *114*, 1343–1360. [CrossRef]
75. Clisby, N.; McCoy, B. Negative Virial Coefficients and the Dominance of Loose Packed Diagrams for D-Dimensional Hard Spheres. *J. Stat. Phys.* **2004**, *114*, 1361–1392. [CrossRef]
76. Bishop, M.; Masters, A.; Vlasov, A.Y. The eighth virial coefficient of four- and five-dimensional hard hyperspheres. *J. Chem. Phys.* **2005**, *122*, 154502. [CrossRef] [PubMed]
77. Clisby, N.; McCoy, B.M. New results for virial coeffcients of hard spheres in D dimensions. *Pramana* **2005**, *64*, 775–783. [CrossRef]
78. Lyberg, I. The fourth virial coefficient of a fluid of hard spheres in odd dimensions. *J. Stat. Phys.* **2005**, *119*, 747–764. [CrossRef]
79. Clisby, N.; McCoy, B.M. Ninth and Tenth Order Virial Coefficients for Hard Spheres in D Dimensions. *J. Stat. Phys.* **2006**, *122*, 15–57. [CrossRef]
80. Zhang, C.; Pettitt, B.M. Computation of high-order virial coefficients in high-dimensional hard-sphere fluids by Mayer sampling. *Mol. Phys.* **2016**, *112*, 1427–1447. [CrossRef]
81. Skoge, M.; Donev, A.; Stillinger, F.H.; Torquato, S. Packing Hyperspheres in high-dimensional Euclidean spaces. *Phys. Rev. E* **2006**, *74*, 041127. [CrossRef]
82. Torquato, S.; Stillinger, F.H. New Conjectural Lower Bounds on the Optimal Density of Sphere Packings. *Exp. Math.* **2006**, *15*, 307–331. [CrossRef]
83. Torquato, S.; Stillinger, F.H. Exactly Solvable Disordered Hard-Sphere Packing Model in Arbitrary-Dimensional Euclidean Spaces. *Phys. Rev. E* **2006**, *73*, 031106. [CrossRef] [PubMed]
84. Torquato, S.; Uche, O.U.; Stillinger, F.H. Random sequential addition of hard spheres in high Euclidean dimensions. *Phys. Rev. E* **2006**, *74*, 061308. [CrossRef]
85. Parisi, G.; Zamponi, F. Amorphous packings of hard spheres for large space dimension. *J. Stat. Mech.* **2006**, P03017. [CrossRef]
86. Scardicchio, A.; Stillinger, F.H.; Torquato, S. Estimates of the optimal density of sphere packings in high dimensions. *J. Math. Phys.* **2008**, *49*, 043301. [CrossRef]
87. van Meel, J.A.; Frenkel, D.; Charbonneau, P. Geometrical frustration: A study of four-dimensional hard spheres. *Phys. Rev. E* **2009**, *79*, 030201(R). [CrossRef] [PubMed]
88. Agapie, S.C.; Whitlock, P.A. Random packing of hyperspheres and Marsaglia's parking lot test. *Monte Carlo Methods Appl.* **2010**, *16*, 197–209. [CrossRef]

89. Torquato, S.; Stillinger, F.H. Jammed hard-particle packings: From Kepler to Bernal and beyond. *Rev. Mod. Phys.* **2010**, *82*, 2633–2672. [CrossRef]
90. Zhang, G.; Torquato, S. Precise algorithm to generate random sequential addition of hard hyperspheres at saturation. *Phys. Rev. E* **2013**, *88*, 053312. [CrossRef]
91. Kazav, E.; Berdichevsky, R.; Schwartz, M. Random close packing from hard-sphere Percus-Yevick theory. *Phys. Rev. E* **2019**, *99*, 012146. [CrossRef] [PubMed]
92. Berthier, L.; Charbonneau, P.; Kundu, J. Bypassing sluggishness: SWAP algorithm and glassiness in high dimensions. *Phys. Rev. E* **2019**, *99*, 031301(R). [CrossRef] [PubMed]
93. Santos, A.; Yuste, S.B.; López de Haro, M.; Odriozola, G.; Ogarko, V. Simple effective rule to estimate the jamming packing fraction of polydisperse hard spheres. *Phys. Rev. E* **2014**, *89*, 040302(R). [CrossRef]
94. Bishop, M.; Michels, J.P.J.; de Schepper, I.M. The short-time behavior of the velocity autocorrelation function of smooth, hard hyperspheres in three, four and five dimensions. *Phys. Lett. A* **1985**, *111*, 169–170. [CrossRef]
95. Colot, J.L.; Baus, M. The freezing of hard disks and hyperspheres. *Phys. Lett. A* **1986**, *119*, 135–139. [CrossRef]
96. Lue, L. Collision statistics, thermodynamics, and transport coefficients of hard hyperspheres in three, four, and five dimensions. *J. Chem. Phys.* **2005**, *122*, 044513. [CrossRef]
97. Santos, A. Note: An exact scaling relation for truncatable free energies of polydisperse hard-sphere mixtures. *J. Chem. Phys.* **2012**, *136*, 136102. [CrossRef]
98. Santos, A. Class of consistent fundamental-measure free energies for hard-sphere mixtures. *Phys. Rev. E* **2012**, *86*, 040102(R). [CrossRef]

© 2020 by the authors. Licensee MDPI, Basel, Switzerland. This article is an open access article distributed under the terms and conditions of the Creative Commons Attribution (CC BY) license (http://creativecommons.org/licenses/by/4.0/).

Article

Time Evolution Features of Entropy Generation Rate in Turbulent Rayleigh-Bénard Convection with Mixed Insulating and Conducting Boundary Conditions

Yikun Wei [1], Pingping Shen [1], Zhengdao Wang [1,*,†], Hong Liang [2,*,†] and Yuehong Qian [3]

[1] Joint Engineering Lab of Fluid Transmission System Technology, Faculty of Mechanical Engineering and Automation, Zhejiang Sci-Tech University, Hangzhou 310018, China; yikunwei@zstu.edu.cn (Y.W.); shenpp@cjlu.edu.cn (P.S.)
[2] Department of Physics, Hangzhou Dianzi University, Hangzhou 310018, China
[3] School of Mathematical Science, Soochow University, Suzhou 215006, China; yuehongqian@suda.edu.cn
* Correspondence: dao@zstu.edu.cn (Z.W.); lianghongstefanie@hdu.edu.cn (H.L.)
† These authors contributed equally to this work.

Received: 6 May 2020; Accepted: 7 June 2020; Published: 17 June 2020

Abstract: Time evolution features of kinetic and thermal entropy generation rates in turbulent Rayleigh-Bénard (RB) convection with mixed insulating and conducting boundary conditions at $Ra = 10^9$ are numerically investigated using the lattice Boltzmann method. The state of flow gradually develops from laminar flow to full turbulent thermal convection motion, and further evolves from full turbulent thermal convection to dissipation flow in the process of turbulent energy transfer. It was seen that the viscous, thermal, and total entropy generation rates gradually increase in wide range of $t/\tau < 32$ with temporal evolution. However, the viscous, thermal, and total entropy generation rates evidently decrease at time $t/\tau = 64$ compared to that of early time. The probability density function distributions, spatial-temporal features of the viscous, thermal, and total entropy generation rates in the closed system provide significant physical insight into the process of the energy injection, the kinetic energy, the kinetic energy transfer, the thermal energy transfer, the viscous dissipated flow and thermal dissipation.

Keywords: entropy generation rate; thermal plume; mixed boundary conditions; heat transfer

1. Introduction

The Rayleigh–Bénard (RB) convection is one of most classical natural convections [1–5], which widely occur in a range of natural and industrial applications [1,2], such as in the Earth's core and mantle, atmosphere, oceans and stars, nuclear reactors, crystallization processes, solar heating devices and so on. The RB convection has been extensively investigated by several experiments in the last few decades [6,7], mostly in slender cells of aspect ratio smaller than or equal to unity in order to reach the largest possible Rayleigh numbers (Ra) or to reveal the detailed characteristic mechanisms of turbulent viscous dynamic and heat transport near the walls or central domains [3,4]. The detailed dynamical and statistical insights of the included turbulent transport and their coherent structures also have been increasingly studied in detail by direct numerical simulations (DNS) [8,9].

It is well known that when the DNS of involving no parametrization of subgrid-scale is carried out, all the dynamically important scales are resolved to faithfully represent the flow. Bailon et al. [10] and Schell et al. [11] reported the derived resolution criteria, thus starting the pioneering work [10,11]; subsequent refinements of this criterion were studied [12,13] and the fine boundary layer dynamics resolution were the main focus [14]. It is only recently that the focus of DNS investigations was presented by the bulk of the research in a convection cell with detailed discussions of the scaling statistics and properties of the dissipation fields [15].

In order to understand the global flow and heat transport loss mechanisms in detail, we will present the detailed statistical characteristic mechanisms of the velocity and temperature gradient fields, in particular the related viscous and thermal local entropy generation rates. The viscous and thermal local entropy generation rates as a criterion are used to provide insight into the local viscous and thermal flow loss in the flow field [16–30]. The viscous and thermal components of the local entropy generation rate are derived in the two-dimensional Cartesian space [30]. Their expressions are as follows, respectively [20,28].

$$\dot{S}_u = \frac{\mu}{\theta}\left\{2\left[\left(\frac{\partial u}{\partial x}\right)^2 + \left(\frac{\partial v}{\partial y}\right)^2\right] + \left(\frac{\partial u}{\partial y} + \frac{\partial v}{\partial x}\right)^2\right\} \tag{1}$$

$$\dot{S}_\theta = \frac{k}{\theta^2}\left[\left(\frac{\partial \theta}{\partial x}\right)^2 + \left(\frac{\partial \theta}{\partial y}\right)^2\right] \tag{2}$$

where μ is the dynamics viscosity of fluid, k denotes the thermal conductivity, θ is the temperature, u represents the x-direction velocity and v is the y-direction velocity, respectively. The total entropy generation rate is the summation of the viscous and thermal entropy generation rates, its expression is as follows [22]:

$$\dot{S} = \dot{S}_u + \dot{S}_\theta \tag{3}$$

The Bejan number (Be) is regarded as an effective approach to judge the importance of heat transfer irreversibility in the domain [23]. Rejane et al. proposed the contribution of heat transfer entropy generation on over all entropy generation by using the Be [27]. The Be is defined by the following equation [29]:

$$Be = \frac{\dot{S}_\theta}{\dot{S}} \tag{4}$$

The range of Be is from 0 to 1. When Be is equal to 0, the irreversibility is dominated by fluid friction. Correspondingly, the irreversibility is dominated by heat transfer when the Be is equal to 1. The irreversibility due to heat transfer dominates in the flow when the Be is greater than 1/2. Correspondingly, $Be < 1/2$ implies that the irreversibilities due to the viscous effects dominate the processes. Meanwhile, it is also noted that the heat transfer and fluid friction entropy generation are equal in $Be = 0.5$ [27].

A wide variety of thermal plumes caused by the buoyancy in turbulent RB convection with mixed insulating and conducting boundary conditions play dominant role in the heat transfer. Once time evolution of the heat transfer has been described and understood in classical turbulent RB convection, the time evolution features of kinetic and thermal entropy generation rates in turbulent RB convection with mixed insulating and conducting boundary conditions still be further expanded. How does the mixed insulating and conducting boundary conditions reduce or improve the time evolution features of entropy generation rate? The mixed insulating and conducting boundary conditions considerably affect the time evolution characteristics of thermal plumes and entropy generation rate.

Based on above discussions, our work mainly focuses on the effect of the mixed insulating and conducting boundary conditions on the time evolution features of thermal plumes, the viscous, thermal and total entropy generation rates. The physical insight features of kinetic and thermal entropy generation rates with time evolution are discussed in in turbulent RB convection with the mixed insulating and conducting boundary conditions, which mainly tried to understand the dynamics of fluid. The remainder of this paper is divided into the following parts. In Section 2, the thermal fluid dynamics equations and numerical method will be briefly depicted. In Section 3, the detailed results of numerical simulation and some discussions are presented. Finally, some conclusions are addressed.

2. Convection Diffusion Equation and Numerical Method

In this section, the convection diffusion equation of thermal fluid, and numerical method of solving them are represented, respectively.

2.1. Convection Diffusion Equation of Thermal Fluid

The convection diffusion equation of thermal fluid is the classical Oberbeck-Boussinesq equations [2,3], and their expressions can be given as follows [6,8].

$$\frac{\partial \rho}{\partial t} + \nabla \cdot (\rho \mathbf{u}) = 0 \tag{5}$$

$$\frac{\partial (\rho \mathbf{u})}{\partial t} + \mathbf{u} \cdot \nabla (\rho \mathbf{u}) = -\nabla p + \nabla \cdot (2\rho v \mathbf{S}) - g\beta \Delta \theta \tag{6}$$

$$\frac{\partial \theta}{\partial t} + \mathbf{u} \cdot \nabla \theta = \kappa \nabla^2 \theta \tag{7}$$

in which ρ is the fluid density, $\mathbf{u}$ represents the macroscopic velocity, v is the kinematic viscosity, p is the pressure of fluid, κ denotes the diffusivity, β is the thermal diffusion coefficient, g is the force of gravity, $\Delta \theta$ represents the difference of temperature and θ denotes macroscopic temperature of fluid.

A large number of numerical methods are widely used to solve the classical Oberbeck-Boussinesq equations [31–36]. The finite element methods [36], finite difference method [34] and the finite volume method [35] are traditional macroscopic methods for Computational Fluid Dynamics (CFD) calculation. The lattice Boltzmann method (LBM) is a computational fluid dynamics method based on mesoscopic simulation scale [37–41]. Compared with other traditional CFD calculation methods, this method has mesoscopic model characteristics between a micromolecular dynamics model and a macrocontinuous model. LBM also has the advantage of a simple description of fluid interaction, and is easier to set a complex boundary, reach a parallel calculation, and implement a program, etc. [42]. LBM has been widely considered as an effective method to describe fluid motion and deal with engineering problems [43,44]. In the subsection, double distribution LBM for will be introduced.

2.2. Numerical Method for Convection Diffusion Equation of Thermal Fluid

In the present paper, the double distributions of LBM are implemented to study the convection diffusion equation of thermal fluid, respectively [38,39]. A lattice Boltzmann equation is implemented to simulate the fluid flow field. Its expression is as follows [42]:

$$f_i(\mathbf{x} + \mathbf{c}_i \Delta t, t + \Delta t) = f_i(\mathbf{x}, t) + (f_i^{eq}(\mathbf{x}, t)) - f_i(\mathbf{x}, t))/\tau_v + F_i \tag{8}$$

where $f_i(\mathbf{x}, t)$ denotes the density distribution functions at $(\mathbf{x}, t)$, $\mathbf{c}_i$ represents the discrete velocity. F_i is the discrete force term, $f_i^{eq}(\mathbf{x}, t)$ is the equilibrium function of density distribution, and τ_v denotes the relaxation time. The equilibrium function for the density is given as:

$$f_i^{eq} = \rho w_i [1 + \frac{\mathbf{c}_i \cdot \mathbf{u}}{c_s^2} + \frac{(\mathbf{c}_i \cdot \mathbf{u})^2}{c_s^2} - \frac{u^2}{2c_s^2}] \tag{9}$$

where w_i denotes the weight coefficient [42]. The kinematic viscosity v is computed by the following equation

$$v = \frac{2\tau_v - 1}{6} \frac{(\Delta x)^2}{\Delta t} \tag{10}$$

The lattice Boltzmann equation for the temperature field is given by the following equation

$$g_i(\mathbf{x} + \mathbf{c}_i \Delta t, t + \Delta t) = g_i(\mathbf{x}, t) + (g_i^{eq}(\mathbf{x}, t)) - g_i(\mathbf{x}, t))/\tau_\theta \tag{11}$$

where $g_i(\mathbf{x}, t)$ is the temperature distribution function at $(\mathbf{x}, t)$, τ_θ denotes the relaxation times for temperature evolution equation. g_i^{eq} is the equilibrium function for temperature distribution. Its expression is given as [40]:

$$g_i^{eq} = \theta w_i [1 + \frac{\mathbf{c}_i \cdot \mathbf{u}}{c_s^2} + \frac{(\mathbf{c}_i \cdot \mathbf{u})^2}{c_s^4} - \frac{u^2}{2c_s^2}] \qquad (12)$$

The diffusivity κ is as follows:

$$\kappa = \frac{2\tau_\theta - 1}{6} \frac{(\Delta x)^2}{\Delta t} \qquad (13)$$

The density, macroscopic velocity, and temperature are as follows:

$$\rho = \sum_{i=0}^{8} f_i, \quad \rho \mathbf{u} = \sum_{i=0}^{8} \mathbf{c}_i f_i, \quad \theta = \sum_{i=0}^{8} g_i \qquad (14)$$

The Chapman–Enskog expansions of Equations (8) and (11) are used to derive the classical Oberbeck-Boussinesq equations [42]. A macroscopic length scale ($x_1 = \varepsilon x$) and two macroscopic time scales ($t_1 = \varepsilon t$, $t_2 = \varepsilon t$) are implemented in the Chapman–Enskog expansion. Two time scales $\partial t = \varepsilon \partial_{t1} + \varepsilon^2 \partial_{t2}$ and one spatial scale $\partial_x = \varepsilon \partial_\alpha$ are used for the Frisch, Hasslacher, and Pomeau (FHP) model [38]. The classical Oberbeck-Boussinesq equations can be derived by executing the streaming step and using the above Chapman-Enskog expansion [42].

Figure 1 shows the computational model of geometrical schematic. As shown in Figure 1, the inhomogeneities heat plates are restricted only in the bottom condition ($y = 0$), and are made of alternating regions of either the isothermal boundary condition, $\theta = \theta_{\text{down}}$, where the discrete black region denotes heat source, or adiabatic boundary condition, $\partial_y \theta = 0$. The upper boundary keeps a constant temperature, $\theta = \theta_{\text{up}}$. In this physical model, the ratio of width of dividing height, $\xi = H/L$, and another two nondimensionless parameters are implemented to define the geometrical configuration of the discrete heat source; the ratio of single heat source is defined as $\lambda = l/L$, and the total ratio of discrete heat source area, $\eta = nl/L$, in which n denotes the heat source number and l represents the single heat-source length. When η is equal to 1, the model becomes the classical RB convection. The above several nondimensionless parameters are implemented to obtain a better understanding of heat transfer transport in turbulent Rayleigh-Bénard convection with mixed insulating and conducting boundary conditions.

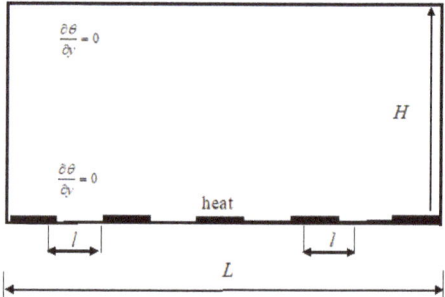

Figure 1. Computational geometry and boundary conditions in the two-dimensional space.

The *Rayleigh* number is one of the most important dimensionless parameter in RB convection. Its expression is as follows:

$$Ra = \beta \Delta \theta g H^3 / \nu \kappa \qquad (15)$$

The *Nusselt* number is one of the most important dimensionless parameter in RB convection to reflect the performance of heat transfer system. It is obtained by:

$$Nu = 1 + \langle u_y \theta \rangle / \kappa \Delta \theta H \tag{16}$$

where $\Delta\theta$ is the temperature difference between the top boundary and the bottom boundary, H denotes the channel height, u_y is the y-velocity, and $\langle \cdot \rangle$ represents the average value of entire domain.

The boundary conditions play a key role in the computational stability and accuracy. The periodic boundary condition and approach of nonequilibrium extrapolation are carried out in this paper. Their ideas will be introduced, respectively. The idea of the periodic boundary condition approach is as follows [42]:

$$f_i(\mathbf{x}, t) = f_i(\mathbf{x} + \mathbf{L}, t) \tag{17}$$

$$g_i(\mathbf{x}, t) = g_i(\mathbf{x} + \mathbf{L}, t) \tag{18}$$

in which the vector $\mathbf{L}$ represents length of the flow pattern. The approach of nonequilibrium extrapolation is as follows [42]:

$$f_i(\mathbf{x}_b, t) = f_i^{eq}(\rho_w, \mathbf{u}_w) + (f_i(\mathbf{x}_f, t) - f_i^{eq}(\rho_f, \mathbf{u}_f)) \tag{19}$$

$$g_i(\mathbf{x}_b, t) = g_i^{eq}(\rho_w, \mathbf{u}_w) + (g_i(\mathbf{x}_f, t) - g_i^{eq}(\rho_f, \mathbf{u}_f)) \tag{20}$$

in which the nonequilibrium contribution can be derived from the fluid node $\mathbf{x}_f$ next to $\mathbf{x}_b$ along the boundary normal vector. During propagation, the unknown incoming populations can be obtained by leaving the domain at the opposite side.

As illustrated in Figure 1, the inhomogeneities heat plates are implemented in all numerical simulations. For $Ra = 10^9$, 4000×2000 lattices in two-dimensional space are implemented to study the temperature fields, viscous, thermal and total entropy generation rates. The parameter λ is equal to 1/9, η is equal to 5/9, the nonequilibrium extrapolation is applied at the top and bottom boundary conditions, the periodic boundary condition is used at left and right boundaries, and the *Prandtl* number ($Pr = \nu/\kappa$) is equal to 1. The dimensionless temperature of discrete heat source equals to 300 in Figure 1, and the dimensionless initial temperature of the fluid is 299.

3. Results and Discussions

The analysis of temperature field, flow streamlines, and various entropy generation rates will be discussed with spatial-temporal evolution in this section, respectively.

3.1. Analysis of Flow and Temperature Field

Figure 2 describes the isotherms' temperature distributions with time evolution at $t/\tau = 8$, $t/\tau = 16$, $t/\tau = 32$, and $t/\tau = 64$. Here, τ ($\tau = \sqrt{H/\beta g \Delta \theta}$) is the characteristic time of the computing system. As described in Figure 2, it can be seen that a few thermal plumes ascend in the region of dimensionless bottom boundary (0.5), two big thermal plumes descend in the region of dimensionless top boundary (1.5) at time $t/\tau = 8$, and the large-scale thermal plumes descend in the region of dimensionless top boundary (0.5) and ascend in the region of dimensionless bottom boundary (1.5) at time $t/\tau = 16$. According to the development phenomenon of thermal plumes at times $t/\tau = 8$ and $t/\tau = 16$, the thermal convective motion of the whole field is still in the initial stage of turbulent development. It was seen that with time evolution, a large-scale thermal plume ascends, strikes on the top plate, and in-volutes several thermal plumes to both sides in the left half of the system, and two large-scale thermal plumes descend at $t/\tau = 32$. These thermal plumes interact and strike on the top and bottom plates with time evolution, a number of small-scale thermal plumes appear at $t/\tau = 32$, which demonstrates that the physical system of thermal convection gradually evolves from the large-scale to small-scale thermal plumes with time evolution [3]. In the process of energy cascade of turbulent thermal convection, the energy of the first large vortex comes from the thermal buoyancy of the outside world, which produces the second small vortex. After the small vortex loses its stability, it produces a smaller vortex process [5]. At $t/\tau = 64$, some smallest plumes can be coagulated into the big plumes, several big plumes reappear with time evolution again. The above phenomenon of temperature distributions with time evolution is consistent with the previous studies [11].

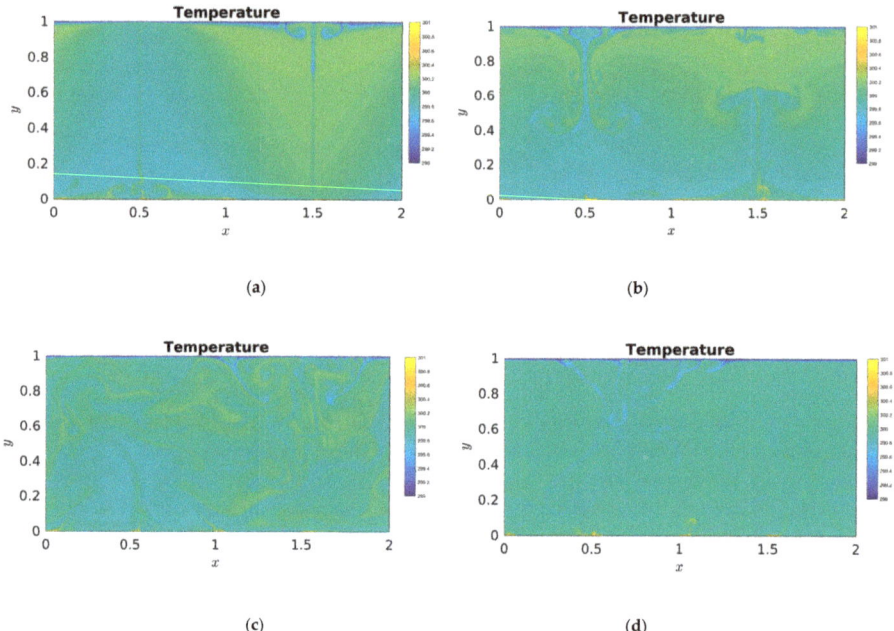

Figure 2. Temperature distributions (isotherms) with time evolution in the two-dimensional RB convection (**a**) $t/\tau = 8$ (**b**) $t/\tau = 16$, (**c**) $t/\tau = 32$, and (**d**) $t/\tau = 64$.

To further demonstrate the above thermal convection flow phenomenon of the whole field, the streamlines of the thermal convection flow at four same time evolution steps are shown in Figure 3. As illustrated in Figure 3, one can clearly see that two large vortexes occur in the central region of the whole field due to the injection of energy at the early characteristic time; two small vortexes appear at dimensionless bottom boundary (0.5) and dimensionless top boundary (1.5) at time $t/\tau = 8$ respectively. In addition, two small vortexes ascend in the region of the dimensionless bottom boundary (0.5), two small vortexes descend in the region of dimensionless top boundary (1.5). At time $t/\tau = 16$, two large vortexes in central region of the whole field become unstable, and more small vortexes appear in the dimensionless top and bottom boundaries (0.5 and 1.5). It was seen that at time $t/\tau = 32$, the early large vortexes evolve into a large number of small scale vortexes, and a large number of small scale vortexes generate due to energy transfer process in the whole field. However, many small vortexes disappear in main flow field, and two relatively big vortexes reappear $t/\tau = 64$. The above phenomenon demonstrates that several large vortexes interact and develop to a large number of small vortexes and a few small vortexes dissipate and big vortexes reappear with temporal evolution, which qualitatively depicts that the state of flow gradually develops from laminar flow to full turbulent thermal convection motion, and further evolve from full turbulent thermal convection to dissipation flow in the process of turbulent energy transfer.

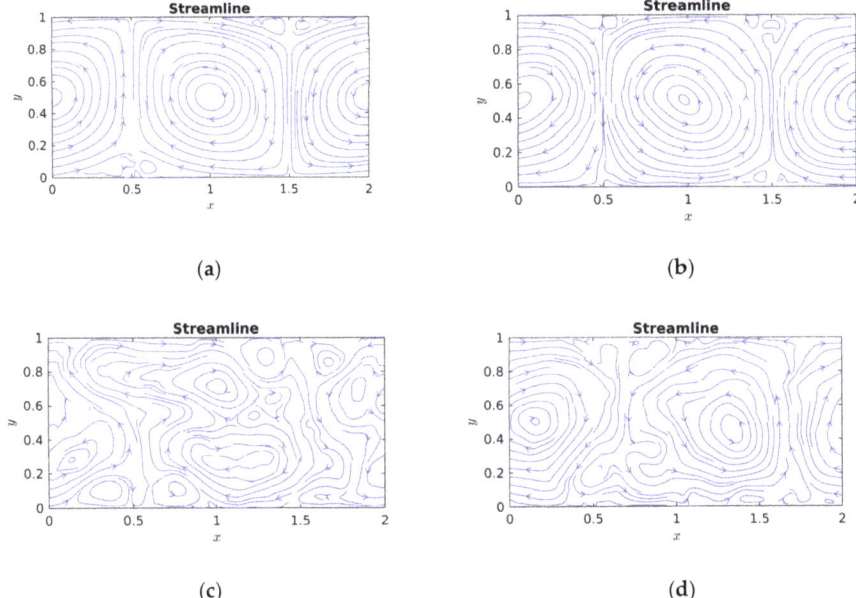

Figure 3. Streamlines of the thermal convection flow with time evolution in the two-dimensional RB convection (**a**) $t/\tau = 8$ (**b**) $t/\tau = 16$, (**c**) $t/\tau = 32$, and (**d**) $t/\tau = 64$.

3.2. Analysis of Entropy Generation Rate

The isotherms temperature distributions and streamlines with time evolution are represented in the above section and several analyses of the entropy generation rate will be discussed in the following section. Figure 4 describes the viscous entropy generation rate at times $t/\tau = 8$, $t/\tau = 16$, $t/\tau = 32$, and $t/\tau = 64$. As shown in Figure 4, it is clearly seen that the high viscous entropy generation rate mainly appears in the intersectional region between the main flow and the top and bottom boundaries and in the intersectional region between big vortexes at times $t/\tau = 8$ and $t/\tau = 16$. It was seen that the high viscous entropy generation rate mainly appears in the high shear region between main flow region and vortex, the low viscous entropy generation rate occurs near the central region of various vortex, which indicates that the viscous flow loss mainly occurs in the high shear region. Meanwhile, at time step $t/\tau = 32$, the viscous entropy generation rate evidently increases with temporal evolution. Nevertheless, the viscous entropy generation rate evidently decreases at time $t/\tau = 64$ compared to that of $t/\tau = 32$, which indicates that the whole mainstream field has already entered the state of turbulent dissipation.

Figure 5 illustrates the thermal entropy generation rate at times $t/\tau = 8$, $t/\tau = 16$, $t/\tau = 32$, and $t/\tau = 64$. Plotted in Figure 5, it is obviously observed that at times $t/\tau = 8$ and $t/\tau = 16$, the high distribution value of thermal entropy generation rate mainly dominates in the high gradient fields of temperature, especially near the top and bottom boundaries compared with the corresponding temperature fields in Figure 2. The low distribution value of thermal entropy generation rate mainly occurs in the homogenetic temperature fields. It is seen that with spatial-temporal evolution, the high distribution value of thermal entropy generation rate gradually increases due to the interaction and strike of these thermal plumes at time $t/\tau = 32$, which indicates that the order degree of thermal movement gradually tends to be disordered in the whole closed system. However, the plume scale of thermal entropy generation rate gradually decreases at time $t/\tau = 64$ compared to that of $t/\tau = 32$, which further demonstrates that a great deal of large scale turbulent structures interact and develop into a large number of small scale turbulent structures; the thermal dissipation also appears with time evolution in the closed system.

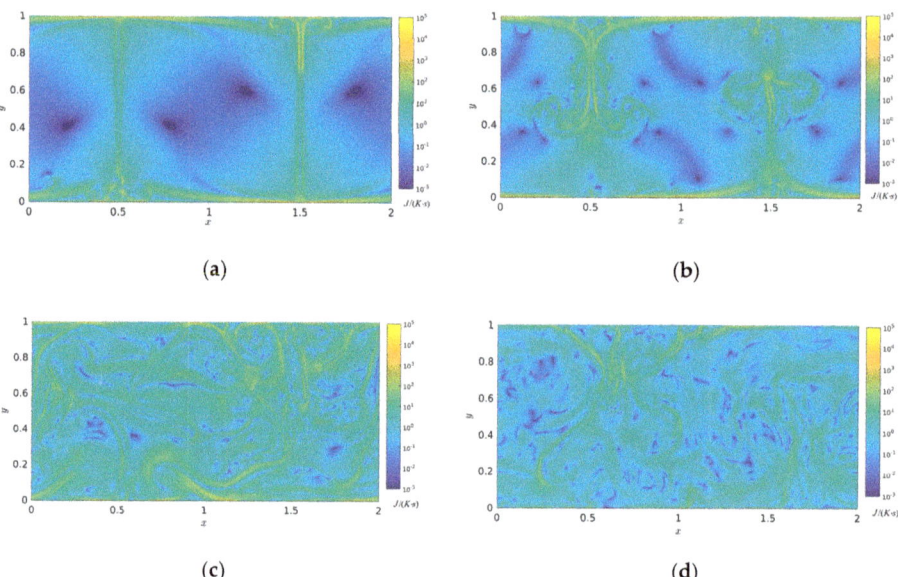

Figure 4. Viscous entropy generation rate with time evolution in the two-dimensional RB convection (**a**) $t/\tau = 8$ (**b**) $t/\tau = 16$, (**c**) $t/\tau = 32$, and (**d**) $t/\tau = 64$ (Units: $J/(K \cdot s)$).

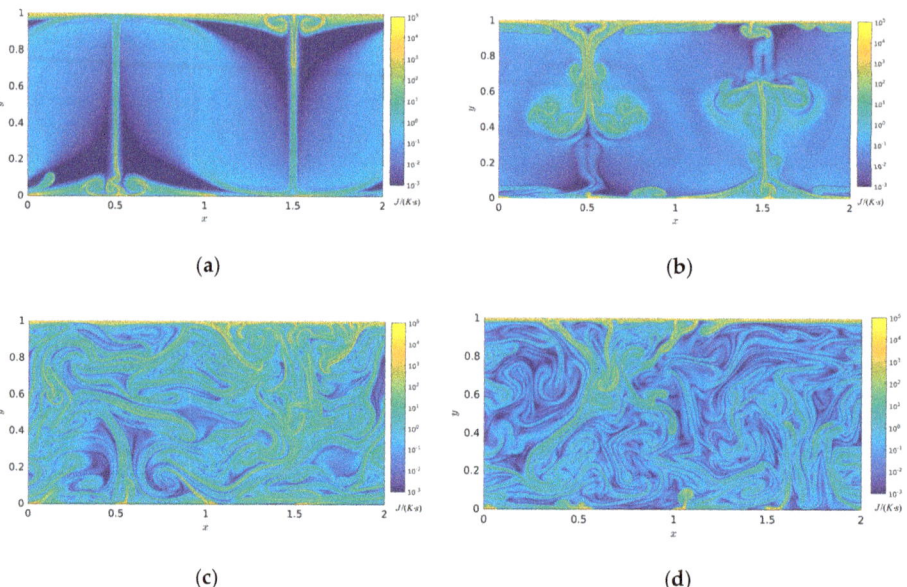

Figure 5. Thermal entropy generation rate with time evolution in the two-dimensional RB convection (**a**) $t/\tau = 8$ (**b**) $t/\tau = 16$, (**c**) $t/\tau = 32$, and (**d**) $t/\tau = 64$ (Units: $J/(K \cdot s)$).

Figure 6 describes the total entropy generation rate at times $t/\tau = 8$, $t/\tau = 16$, $t/\tau = 32$, and $t/\tau = 64$. As described in Figure 6, it can be seen that the high distribution value of total entropy generation rate mainly dominates in the largest temperature velocity gradient compared with the corresponding temperature fields in Figure 2. The low total entropy generation rate mainly clusters in the region of the homogenetic temperature fields. The distribution size trend of total entropy generation rate is well consistent with that of thermal entropy generation rate in the corresponding time step. In the spatial evolution, the shape of high entropy generation rate congeals into a large number of varied plumes, which indicates that the role of thermal entropy generation rate gradually improves with time evolution in the heat transfer irreversibility. It can be clearly seen that with time evolution, a great deal of large scale plumes interact and develop to a large number of small scale plumes in the closed system, and the value of total entropy generation rate increases, which indicates that the order degree of energy dissipation in the whole closed system gradually tends to be disordered and increase. The viscous, thermal and total entropy generation rates with evolution can promote the idea that the type of mixed bottom boundary condition and thermal configuration can be extensively applied in a wide variety of practical engineering applications, such as the solar thermal absorber plate or the electronic existing plates.

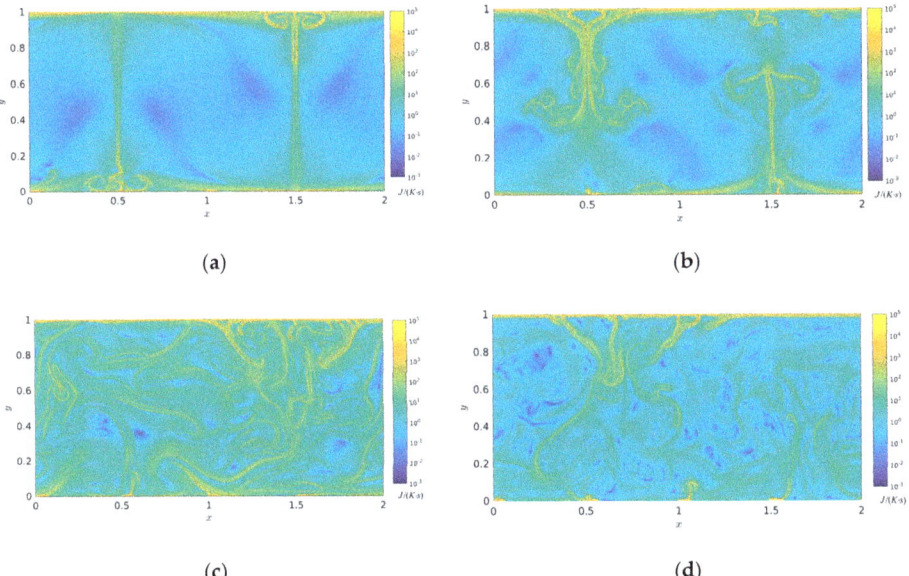

Figure 6. Total entropy generation rate with time evolution in the two-dimensional RB convection (**a**) $t/\tau = 8$ (**b**) $t/\tau = 16$, (**c**) $t/\tau = 32$, and (**d**) $t/\tau = 64$ (Units: $J/(K \cdot s)$).

3.3. Quantitative Analysis of Entropy Generation Rate with Time Evolution

The probability density function (PDF) is used to reveal the distribution aggregation situation of physics variable. Wei et al. [24] argued that the PDFs of S_u, S_θ and S with increase of Prandtl number, the tails of high entropy generation rates can fit well into the curve of the log-normal coordinate and the departure and the distribution of log-normality, gradually becoming more robust with the decrease of Prandtl number. In this paper, an exponential expression is implemented for PDF. Its exponential expression is as follows [24]:

$$p(Y) = \frac{C}{\sqrt{Y}} \exp(-mY^\alpha) \qquad (21)$$

in which m, α and C represent the fitted parameters, and $Y = X - X_{mp}$ with $X = \dot{S}_u/(\dot{S}_u)_{rms}$, $\dot{S}_\theta/(\dot{S}_\theta)_{rms}$, $\dot{S}/(\dot{S})_{rms}$ and X_{mp} being the abscissa of the most probable amplitude. The best fit of Equation (21) to the data yields $m = 0.86$ and $\alpha = 0.72$ for $\dot{S}_u$, $m = 1.15$ and $\alpha = 0.69$ for $\dot{S}_\theta$ and $m = 1.06$ and $\alpha = 0.72$ for $\dot{S}$.

To highlight the distribution aggregation differences of $\dot{S}_u$, $\dot{S}_\theta$ and $\dot{S}$ with time evolution, the PDFs of $\dot{S}_u$, $\dot{S}_\theta$ and $\dot{S}_0$ are plotted, respectively, where $\dot{S}_u$, $\dot{S}_\theta$ and $\dot{S}_0$ represent the value distributions of $\dot{S}_u$, $\dot{S}_\theta$ and $\dot{S}$ in the whole region. Figure 7 describes the PDFs' distributions of $\dot{S}_u$ at four times $t/\tau = 8$, $t/\tau = 16$, $t/\tau = 32$ and $t/\tau = 64$. As described in Figure 7, we can see that the high value of $\dot{S}_u$ decreases in a range of $\dot{S}_u > 10$ with time evolution. This is mainly due to the fact that the flow characteristic velocity of the large-scale flow in early characteristic time at $t/\tau = 8$ is relatively large, the large-scale flow is broken into more small-scale flows, the viscosity entropy rate decreases in high value with time evolution, and the viscosity entropy generation rate of the small-scale flow is smaller than that of the large-scale flow.

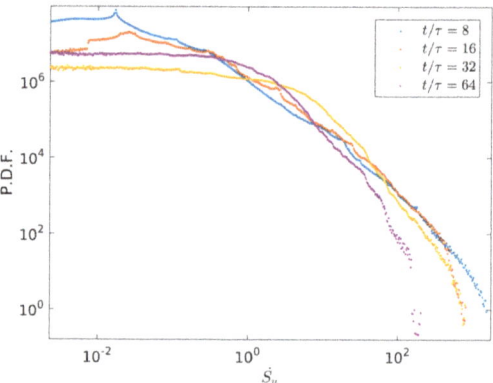

Figure 7. PDF distributions of viscous entropy generation rate in the two-dimensional RB convection and at four times $t/\tau = 8$, $t/\tau = 16$, $t/\tau = 32$, and $t/\tau = 64$.

Figure 8 shows the PDF distributions of thermal entropy generation rate at four times $t/\tau = 8$, $t/\tau = 16$, $t/\tau = 32$, and $t/\tau = 64$. Plotted in Figure 8, it is clearly obtained that the high values of $\dot{S}_\theta$ keeps almost the same in a wide range of $\dot{S}_u > 100$ with time evolution, the low and middle values of $\dot{S}_\theta$ keep light difference in a wide range of $\dot{S}_u < 100$ with time evolution. Figure 9 illustrates the PDF distributions of total entropy generation rate $\dot{S}_0$ at four times $t/\tau = 8$, $t/\tau = 16$, $t/\tau = 32$, and $t/\tau = 64$. As illustrated in Figure 9, it can be seen that the high values of $\dot{S}_0$ keep almost the same with $\dot{S}_\theta$, which indicates that the thermal entropy generation rate has a dominant position in the total entropy generation rate with time evolution.

To further reveal the distribution differences of $\dot{S}_u$, $\dot{S}_\theta$ and $\dot{S}$ with time evolution, the average value of $\dot{S}_u$, $\dot{S}_\theta$ and $\dot{S}_0$ are plotted in the whole region, respectively. $\overline{\dot{S}_u}$, $\overline{\dot{S}_\theta}$, and $\overline{\dot{S}_0}$ denote the average value of $\dot{S}_u$, $\dot{S}_\theta$ and $\dot{S}_0$ in the whole region. Figure 10 shows the time evolution of average viscous entropy generation rate from $t/\tau = 0$ to $t/\tau = 100$ in the whole field. Plotted in Figure 10, it is clearly observed that the average value of $\dot{S}_u$ alternately increases, three peaks successively appear from the time step of $t/\tau = 0$ to $t/\tau = 32$ with time evolution. One strong peak appears at the time step of $t/\tau = 32$, however, the average value of $\overline{\dot{S}_u}$ gradually decreases from the time step of $t/\tau = 32$ to $t/\tau = 64$, and the average value of $\overline{\dot{S}_u}$ gradually increases in a range of $t/\tau > 64$. This is mainly due to the fact that the largest length-scales eddy is produced owing to the injection of energy at an early characteristic time;

the decrease of flow eddies and the geometric eddy size is associated with the characteristic time scales ($t/\tau < 32$). However, with time evolution, the large-scale flow is broken into more small-scale flows in a range of t/τ from 32 to 64; the viscosity entropy rate decreases in high value with time evolution, and the viscosity entropy generation rate of the small-scale flow is smaller than that of the large-scale flow. In a range of $t/\tau > 64$, some of the smallest eddies can be distorted in this distortion process, which further indicates that the kinetic energy may be dissipated from the dissipation of the smallest eddies owing to the effect of viscous flow.

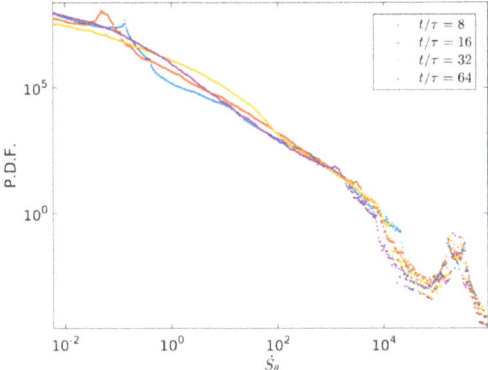

Figure 8. PDF distributions of thermal entropy generation rate in the two-dimensional RB convection and at four times $t/\tau = 8$, $t/\tau = 16$, $t/\tau = 32$, and $t/\tau = 64$.

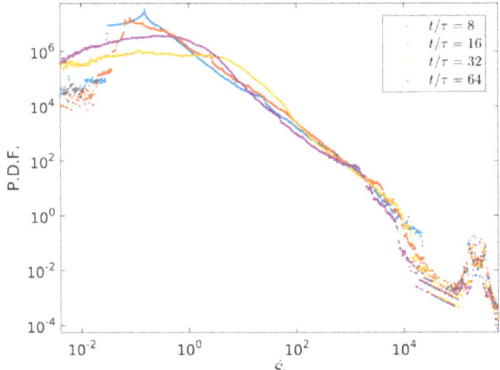

Figure 9. PDF distributions of total entropy generation rate in the two-dimensional RB convection and at four times $t/\tau = 8$, $t/\tau = 16$, $t/\tau = 32$, and $t/\tau = 64$.

Figure 11 illustrates the time evolution of average thermal entropy generation rate from $t/\tau = 0$ to $t/\tau = 100$ in the whole field. As illustrated in Figure 11, one can clearly see that at first $\overline{\dot{S}_\theta}$ is very large due to the extremely thin boundary layer. As time goes by, the boundary layer thickness rapidly increases to the normal level, and $\overline{\dot{S}_\theta}$ decreases rapidly. After the initial period, the average value of the temperature generation rate $\overline{\dot{S}_\theta}$ alternately increases, several peaks periodically appear from the time step of $t/\tau = 0$ to $t/\tau = 32$ with time evolution. One strong peak appears at the time step of $t/\tau = 32$, however, the average value of $\overline{\dot{S}_\theta}$ periodically decreases from the time step of $t/\tau = 32$ to $t/\tau = 64$. The average value of $\overline{\dot{S}_\theta}$ periodically and lightly increases in a range of $t/\tau > 64$. This is mainly due to the fact that the large scale plumes produce owing to the injection of energy at the early characteristic time, the decrease of thermal plumes size is associated with the characteristic time-scales ($t/\tau < 32$). Nevertheless, with time evolution, the large-scale plumes are broken into more small-scale plumes in a range of t/τ from 32 to 64, the thermal entropy rate decreases in high value with time evolution, and the thermal entropy generation rate of the small-scale plumes is smaller than that of the large-scale plumes. In a range of $t/\tau > 64$, some of the smallest plumes can be coagulated into the big plumes.

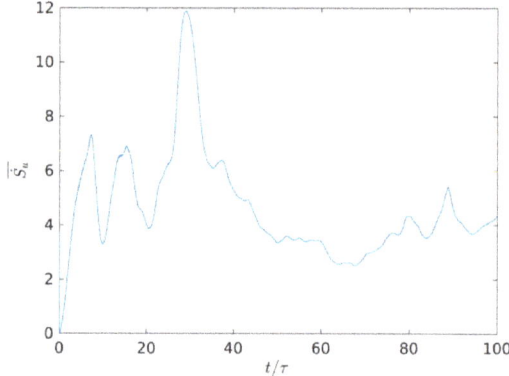

Figure 10. Time evolution of average viscous entropy generation rate in the whole two-dimensional field.

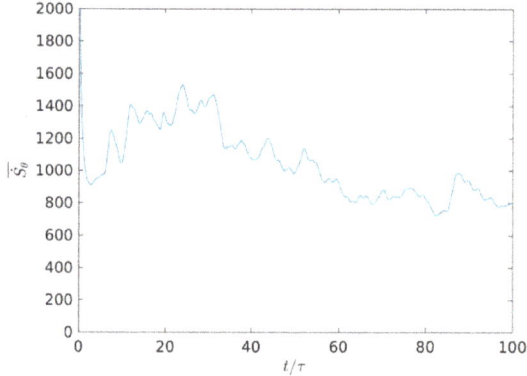

Figure 11. Time evolution of average thermal entropy generation rate in the whole two-dimensional field.

Figure 12 shows the time evolution of average total entropy generation rate from $t/\tau = 0$ to $t/\tau = 100$ in the whole field. As shown in Figure 12, one can clearly see that at first $\bar{S}_0$ is very large due to the extremely thin boundary layer. With time evolution, it is clearly seen that the boundary layer thickness rapidly increases to the normal level, and $\bar{S}_0$ decreases rapidly. Plotted in Figure 12, it can be seen that the high values of $\bar{S}_0$ remain almost the same with $\bar{S}_\theta$ with time evolution, which indicates that the thermal entropy generation rate plays a dominated role in the heat transfer irreversibility—the viscous entropy generation can be neglected time evolution. The above phenomenon is well consistent with the importance of heat transfer irreversibility in the previous studies [23–25]. Wei et al. [24] studied the effect of changing the Prandtl number on the entropy generation rate in two-dimensional RB convection, and argued that the thermal entropy generation rate has a dominant role in the heat transfer irreversibility—the viscous entropy generation can be neglected with the increasing Prandtl number. Mohamed et al. [45–47] studied a new analytical solution of a longitudinal fin with variable heat generation and thermal conductivity in the mixed convection Falkner-Skan flow of nanofluids with variable thermal conductivity.

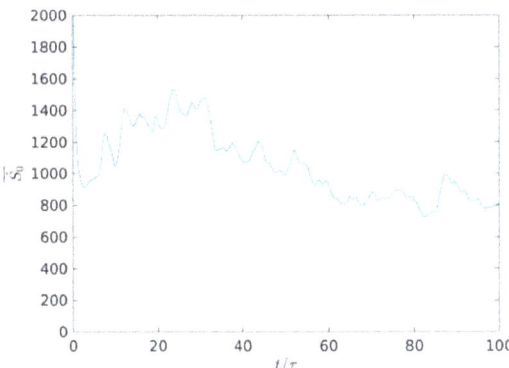

Figure 12. Time evolution of average total entropy generation rate in the whole two-dimensional field.

4. Conclusions

In this paper, the time evolution features of entropy generation rate in turbulent Rayleigh-Bénard convection are investigated in mixed insulating and conducting boundary conditions. Several conclusions are given as follows.

The physical system of thermal convection gradually evolves from the large-scale to small-scale thermal plumes—some of the smallest plumes can be coagulated into the big plumes and several big plumes reappear in the time evolution. The state of flow gradually develops from laminar flow to turbulent thermal convection motion, and further evolves from turbulent thermal convection to dissipation flow in the process of turbulent energy transfer.

The viscous, thermal, and total entropy generation rates evidently increase in wide range of $t/\tau < 32$ with temporal evolution. Nevertheless, the viscous, thermal, and total entropy generation rates evidently decreases at time $t/\tau = 64$ compared to that of $t/\tau = 32$.

The high value of $\bar{S}_u$ decreases in a range of $\bar{S}_u > 10$ with time evolution. It is revealed that the flow characteristic velocity of the large-scale flow in early characteristic time at $t/\tau = 8$ is relatively large: the large-scale flow is broken into more small-scale flows, the viscosity entropy rate decreases in high value with time evolution, and the viscosity entropy generation rate of the small-scale flow is smaller than that of the large-scale flow.

It was seen that the largest length-scale eddy produces owing to the injection of energy at the early characteristic time, the decrease of flow eddies and geometric eddy size are associated with the characteristic time-scales ($t/\tau < 32$). However, the large-scale flow is broken into more small-scale flows in a range of t/τ from 32 to 64, the viscosity entropy rate decreases in high value with time evolution, and the viscosity entropy generation rate of the small-scale flow is smaller than that of the large-scale flow. In a range of $t/\tau > 64$, some of the smallest eddies can be distorted in this distortion process. The average value of $\bar{S}_\theta$ alternately increases from the time step of $t/\tau = 0$ to $t/\tau = 32$, however, the average value of $\bar{S}_\theta$ periodically decreases from the time step of $t/\tau = 32$ to $t/\tau = 64$. Interestingly, the average value of $\bar{S}_\theta$ periodically and lightly increases in a range of $t/\tau > 64$.

The above studies further demonstrate the process of the energy injection, the kinetic energy, the kinetic energy transfer, the thermal energy transfer, the viscous dissipated flow and thermal dissipation. In practical engineering, the type of mixed-bottom boundary condition and thermal configuration can be extensively applied in a wide variety of equipment, such as the solar thermal absorber plate or the electronic existing plates.

Author Contributions: Conceptualization, Y.W. and Y.Q.; Methodology, Z.W.; Software, H.L.; Validation, Z.W., P.S. and Y.Q.; Formal Analysis, P.S.; Investigation, Z.W.; Resources, Y.Q.; Data Curation, Z.W.; Writing—Original Draft Preparation, Z.W.; Writing—Review & Editing, H.L.; Visualization, Y.Q.; Supervision, Y.W.; Project Administration, Y.W.; Funding Acquisition, Y.Q. All authors have read and agreed to the published version of the manuscript.

Funding: This research was funded by the National Natural Science Foundation of China [11872337 and 11902291], the Research Initiation Fund of Zhejiang Sci-Tech University (19022105-Y), the Fundamental Research Funds of Zhejiang Sci-Tech University (2019Y004) and the Dept. Social Security and Human Resources, Project of Ten Thousand Talents, no. W01060069. The referees' valuable suggestions are greatly appreciated.

Conflicts of Interest: The authors declare there are no known conflicts of interest associated with this publication and there has been no significant financial support for this work that could have influenced its outcome. We confirm that the paper has been read and approved by all named authors and that there are no other persons who satisfied the criteria for authorship but are not listed. We further confirm that the order of authors listed in the paper has been approved by all of us. We confirm that we have given due consideration to the protection of intellectual property associated with this work and that there are no impediments to publication, including the timing of publication, with respect to intellectual property. In so doing, we confirm that we have followed the regulations of our institutions concerning intellectual property.

References

1. Lage, J.L.; Lim, J.S.; Bejan, A. Natural convection with radiation in a cavity with open top end. *J. Heat Transf.* **1992**, *114*, 479–486. [CrossRef]
2. Xu, F.; Saha, S.C. Transition to an unsteady flow induced by a fin on the sidewall of a differentially heated air-filled square cavity and heat transfer. *Int. J. Heat Mass Transf.* **2014**, *71*, 236–244. [CrossRef]
3. Nelson, J.E.B.; Balakrishnan, A.R.; Murthy, S.S. Experiments on stratified chilled-water tanks. *Int. J. Refrig.* **1999**, *22*, 216–234. [CrossRef]
4. Ampofo, F.; Karayiannis, T.G. Experimental benchmark data for turbulent natural convection in an air filled square cavity. *Int. J. Heat. Mass Transf.* **2003**, *19*, 3551–3572. [CrossRef]
5. Adeyinka, O.B.; Naterer, G.F. Experimental uncertainty of measured entropy production with pulsed laser PIV and planar laser induced fluorescence. *Appl. Ther. Eng.* **2005**, *48*, 1450–1461. [CrossRef]
6. Lohse, D.; Xia, K.Q. Small-scale properties of turbulent Rayleigh–Bénard convection. *Annu. Rev. Fluid Mech.* **2010**, *42*, 335–364. [CrossRef]
7. Xu, F.; Patterson, J.C.; Lei, C. On the double-layer structure of the thermal boundary layer in a differentially heated cavity. *Int. J. Heat Mass Transf.* **2008**, *51*, 3803–3815. [CrossRef]
8. Ahlers, G.; Grossmann, S.; Lohse, D. Heat transfer and large scale dynamics in turbulent Rayleigh-Bénard convection. *Rev. Mod. Phys.* **2009**, *81*, 503–537. [CrossRef]
9. Sun, C.; Zhou, Q.; Xia, K.Q. Cascades of velocity and temperature fluctuations in buoyancy-driven thermal turbulence. *Phys. Rev. Lett.* **2006**, *97*, 144504–144509. [CrossRef] [PubMed]
10. Bailon-Cuba, J.; Emran, M.S.; Schumacher, J. Aspect ratio dependence of heat transfer and large-scale flow in turbulent convection. *J. Fluid Mech.* **2010**, *655*, 152–173. [CrossRef]
11. Scheel, J.D.; Kim, E.; White, K.R. Thermal and viscous boundary layers in turbulent Rayleigh-Benard convection. *J. Fluid Mech.* **2012**, *711*, 281–305. [CrossRef]
12. Zhou, Q.; Xia, K.Q. Physical and geometrical properties of thermal plumes in turbulent Rayleigh-Bénard convection. *New J. Phys.* **2012**, *12*, 075006–075018. [CrossRef]
13. Shi, N.; Emran, M.S.; Schumacher, J. Boundary layer structure in turbulent Rayleigh-Benard convection. *J. Fluid Mech.* **2012**, *706*, 5–33. [CrossRef]
14. Shishkina, O.; Stevens, R.A.J.M.; Grossmann, S.; Lohse, D. Boundary layer structure in turbulent thermal convection and its consequences for the required numerical resolution. *New J. Phys.* **2010**, *12*, 075022. [CrossRef]
15. Shishkina, O.; Wagner, C. Analysis of sheet like thermal plumes in turbulent Rayleigh-Bénard convection. *J. Fluid Mech.* **2012**, *599*, 383–404. [CrossRef]
16. Lami, P.A.K.; Praka, K.A. A numerical study on natural convection and entropy generation in a porous enclosure with heat sources. *Int. J. Heat Mass Transf.* **2014**, *69*, 390–407. [CrossRef]

17. Zahmatkesh, I. On the importance of thermal boundary conditions in heat transfer and entropy generation for natural convection inside a porous enclosure. *Int. J. Therm. Sci.* **2008**, *47*, 339–346. [CrossRef]
18. Andreozzi, A.; Auletta, A.; Manca, O. Entropy generation in natural convection in a symmetrically and uniformly heated vertical channel. *Int. J. Heat Mass Transf.* **2006**, *49*, 3221–3228. [CrossRef]
19. Dagtekin, I.; Oztop, H.F.; Bahloul, A. Entropy generation for natural convection in Γ-shaped enclosures. *Int. Commun. Heat Mass Transf.* **2007**, *34*, 502–510. [CrossRef]
20. Kaczorowski, M.; Wagner, C. Analysis of the thermal plumes in turbulent Rayleigh-Bénard convection based on well-resolved numerical simulations. *J. Fluid Mech.* **2011**, *618*, 89–112. [CrossRef]
21. Usman, M.; Soomro, F.A.; Haq, R.U.; Wang, W.; Defterli, O. Thermal and velocity slip effects on Casson nanofluid flow over an inclined permeable stretching cylinder via collocation method. *Int. J. Heat Mass Transf.* **2018**, *122*, 1255–1263. [CrossRef]
22. Wang, Z.D.; Qian, Y.H. Numerical study on entropy generation in thermal convection with differentially discrete heat boundary conditions. *Entropy* **2018**, *20*, 351. [CrossRef]
23. Sciacovelli, A.; Verda, V.; Sciubba, E. Entropy generation analysis as a design tool—A review. *Renew. Sustain. Energy Rev.* **2015**, *43*, 1167–1181. [CrossRef]
24. Wei, Y.K.; Wang, Z.D.; Qian, Y.H. A numerical study on entropy generation in two-dimensional Rayleigh-Bénard convection at different Prandtl number. *Entropy* **2017**, *19*, 443. [CrossRef]
25. Jin, Y. Second-law analysis: A powerful tool for analyzing Computational Fluid Dynamics results. *Entropy* **2017**, *19*, 679. [CrossRef]
26. Pizzolato, A.; Sciacovelli, A.; Verda, V. Transient local entropy generation analysis for the design improvement of a thermocline thermal energy storage. *Appl. Therm. Eng.* **2016**, *101*, 622–629. [CrossRef]
27. Rejane, D.C.; Mario, H.; Copetti, J.B. Entropy generation and natural convection in rectangular cavities. *Appl. Therm. Eng.* **2009**, *29*, 1417–1425.
28. Mahian, O.; Kianifar, A.; Pop, I. A review on entropy generation in nanofluid flow. *Int. J. Heat Mass Transf.* **2013**, *65*, 514–532. [CrossRef]
29. Bhatti, M.M.; Rashidi, M.M. Entropy generation with nonlinear thermal radiation in MHD boundary layer flow over a permeable shrinking/stretching sheet: Numerical solution. *J. Nanofluids* **2016**, *5*, 543–554. [CrossRef]
30. Abbas, M.A.; Bai, Y.; Rashidi, M.M.; Bhatti, M.M. Analysis of Entropy Generation in the Flow of Peristaltic Nanofluids in Channels with Compliant Walls. *Entropy* **2016**, *18*, 90. [CrossRef]
31. Lun, Y.X.; Lin, L.M.; He, H.J.; Zhu, Z.C.; Wei, Y.K. Effects of Vortex Structure on Performance Characteristics of a Multiblade Fan with Inclined tongue. *Proc. Inst. Mech. Eng. Part A J. Power Energy* **2019**, *233*, 1007–1021. [CrossRef]
32. Zheng, X.; Lin, Z.; Xu, B.Y. Thermal conductivity and sorption performance of nano-silver powder/FAPO-34 composite fin. *Appl. Therm. Eng.* **2019**, *160*, 114055–114063. [CrossRef]
33. Lin, Z.; Liu, Z.X.; Liu, Q.; Li, Y. Fluidization characteristics of particles in a groove induced by horizontal air flow. *Powder Technol.* **2020**, *363*, 442–447. [CrossRef]
34. Yang, H.; Yu, P.Q.; Xu, J.; Zhu, Z.C. Experimental investigations on the performance and noise characteristics of a forward-curved fan with the stepped tongue. *Meas. Control* **2019**, *52*, 1480–1488. [CrossRef]
35. Zhang, W.; Chen, X.P.; Zhu, Z.C. Partitioning effect on natural convection in a circular enclosure with an asymmetrically placed inclined plate. *Int. Commun. Heat Mass Transf.* **2018**, *90*, 11–22. [CrossRef]
36. Xu, H.; Cantwell, C.D.; Monteserin, C.; Eskilsson, A.P.; Engsig-Karup, A.P.; Sherwin, S.J. Spectral/hp element methods: Recent developments, applications, and perspectives. *J. Hydrodyn.* **2018**, *30*, 1–22. [CrossRef]
37. Liu, H.H.; Valocch, A.J.; Zhang, Y.H.; Kang, Q.J. Lattice Boltzmann Phase Field Modeling Thermocapillary Flows in a Confined Microchannel. *J. Comput. Phys.* **2014**, *256*, 334–356. [CrossRef]
38. Shan, X. Simulation of Rayleigh-Bénard convection using a lattice Boltzmann method. *Phys. Rev. E* **1997**, *55*, 2780–2788. [CrossRef]
39. Liang, H.; Xu, J.; Chai, Z.H.; Shi, B.C. Lattice boltzmann modeling of wall-bounded ternary fluid flows. *Appl. Math. Model.* **2019**, *73*, 487–513. [CrossRef]
40. Wei, Y.K.; Wang, Z.D.; Yang, J.F.; Dou, H.S.; Qian, Y.H. Simulation of natural convection heat transfer in an enclosure at different Rayleigh number using lattice Boltzmann method. *Comput. Fluids* **2016**, *124*, 30–38. [CrossRef]

41. Wei, Y.K.; Wang, Z.D.; Dou, H.S.; Qian, Y.H. A novel two-dimensional coupled lattice Boltzmann model for incompressible flow in application of turbulence Rayleigh-Taylor instability. *Comput. Fluids* **2017**, *156*, 97–102. [CrossRef]
42. Chen, S.Y.; Doolen, G.D. Lattice Boltzmann method for fluid flows. *Annu. Rev. Fluid Mech.* **1998**, *30*, 329–364. [CrossRef]
43. Wang, Z.D.; Wei, Y.; Qian, Y.H. A bounce back-immersed boundary-lattice Boltzmann model for curved boundary. *Appl. Math. Model.* **2020**, *81*, 428–440. [CrossRef]
44. Chen, Z.; Shu, C. Simplified lattice Boltzmann method for non-Newtonian power-law fluid flows. *Int. J. Numer. Methods Fluids* **2020**, *92*, 38–54. [CrossRef]
45. Mohamed, K.; Ismai, T.; Mohamed, R.E. A new analytical solution of longitudinal fin with variable heat generation and thermal conductivity using DRA. *Eur. Phys. J. Plus* **2020**, *135*, 120–129.
46. Nawel, B.; Mohamed, K.; Ismai, T.; Mohamed, R.E. On numerical and analytical solutions for mixed convection Falkner-Skan flow of nanofluids with variable thermal conductivity. *Waves Random Complex Media* **2019**, 1–19. [CrossRef]
47. Mohamed, R.E. Effects of NP Shapes on Non-Newtonian Bio-Nanofluid Flow in Suction/Blowing Process with Convective Condition: Sisko Model. *J. Non-Equilib. Thermodyn.* **2020**, *45*, 97–108.

© 2020 by the authors. Licensee MDPI, Basel, Switzerland. This article is an open access article distributed under the terms and conditions of the Creative Commons Attribution (CC BY) license (http://creativecommons.org/licenses/by/4.0/).

Article

Entropy Multiparticle Correlation Expansion for a Crystal

Santi Prestipino * and Paolo V. Giaquinta

Dipartimento di Scienze Matematiche ed Informatiche, Scienze Fisiche e Scienze della Terra, Università degli Studi di Messina, Viale F. Stagno d'Alcontres 31, 98166 Messina, Italy; paolo.giaquinta@unime.it
* Correspondence: sprestipino@unime.it

Received: 21 August 2020; Accepted: 11 September 2020; Published: 13 September 2020

Abstract: As first shown by H. S. Green in 1952, the entropy of a classical fluid of identical particles can be written as a sum of many-particle contributions, each of them being a distinctive functional of all spatial distribution functions up to a given order. By revisiting the combinatorial derivation of the entropy formula, we argue that a similar correlation expansion holds for the entropy of a crystalline system. We discuss how one- and two-body entropies scale with the size of the crystal, and provide fresh numerical data to check the expectation, grounded in theoretical arguments, that both entropies are extensive quantities.

Keywords: entropy multiparticle correlation expansion; one- and two-body density functions; one- and two-body entropy

1. Introduction

The entropy multiparticle correlation expansion (MPCE) is an elegant statistical-mechanical formula that entails the possibility of reconstructing the total entropy of a many-particle system term by term, including at each step of summation the integrated contribution from spatial correlations between a specified number of particles.

The original derivation of the entropy MPCE is found in a book by H. S. Green (1952) [1]. Green's expansion applies for the canonical ensemble (CE). In 1958, Nettleton and M. S. Green derived an apparently different expansion valid in the grand-canonical ensemble (GCE) [2]. It took the ingenuity of Baranyai and Evans to realize, in 1989, that the CE expansion can indeed be reshuffled in such a way as to become formally equivalent to the GCE expansion [3].

A decisive step forward was eventually taken by Schlijper [4] and An [5], who have highlighted the similarity of the entropy formula to a cumulant expansion, and the close relationship with the cluster variation method (see, e.g., [6]). Other papers wherein in various ways the combinatorial content of the entropy MPCE is emphasized are references [7–10].

Since the very beginning it has been clear that the successive terms in the entropy expansion for a homogeneous fluid are not all of equal importance. In particular, the contributions from correlations between more than two particles are only sizable at moderate and higher densities. However, while the two-body entropy is easily accessed in a simulation, computing the higher-order entropy terms is a prohibitive task (see, however, reference [11]). Hence, the only viable method to compute the total entropy in a simulation remains thermodynamic integration (see e.g., [12]). The practical interest for the entropy expansion has thus shifted towards the residual multiparticle entropy (RMPE), defined as the difference between excess entropy and two-body entropy. The RMPE is a measure of the impact of non-pair multiparticle correlations on the entropy of the fluid. For hard spheres, Giaquinta and Giunta have observed that the RMPE changes sign from negative to positive very close to freezing [13]. At low densities the RMPE is negative, reflecting a global reduction (largely driven by two-body correlations) of the phase space available to the system as compared to the ideal gas. The change of sign of the RMPE close to freezing indicates that fluid particles, which at high enough densities are forced by more stringent packing constraints, start exploring, this time in a cooperative way, a different structural condition on a local scale, preluding to crystallization on a global scale. Since the original observation in [13], a clear correspondence between the RMPE zero and the ultimate threshold for spatial homogeneity in the system has been found in many simple and complex fluids [14–24], thereby leading to the belief that the vanishing of the RMPE is a signature of an impending structural or thermodynamic transition of the system from a less ordered to a more spatially organized condition (freezing is just an example of many). Albeit empirical, this entropic criterion is a valid alternative to the far more demanding exact free-energy methods when a rough estimate of the transition point is deemed sufficient. For a simple discussion of the interplay between entropy and ordering, the reader is referred to reference [25]; see instead references [26,27] for general considerations about the entropy of disordered solids.

A pertinent question to ask is, what happens to the RMPE on the solid side of the phase boundary, considering that an entropy expansion also holds for the crystal? This is precisely the problem addressed in this paper. Can the scope of the entropic criterion be extended in such a way that it also applies for melting? As it turns out, we can offer no definite answer to this question, since theory alone does not go far enough and we ran into a serious computational bottleneck: while the formulae are clear and the numerical procedure is straightforward, it is extremely hard to obtain reliable data for the two-body entropy of a three-dimensional crystal. We have only carried out a limited test on a triangular crystal of hard disks, but our results are affected by finite-size artifacts that make them inconclusive. Nevertheless, a few firm points have been established: (1) the approximate entropy expressions obtained by truncating the MPCE at a given order can all be derived from an explicit functional of the correlation functions up to that order; (2) the one-body entropy for a crystal is an extensive quantity (the same is held to be true for the two-body entropy, but our arguments are not sufficient for a proof); (3) the peaks present in the crystal one-body density have a nearly Gaussian shape; (4) we have also clarified the role of lattice symmetries in dictating the structure of the two-body density, which is explicitly determined at zero temperature.

This paper is organized as follows. In Section 2 we resume the formalism of the entropy expansion for homogeneous fluids and provide the basic tools needed for its extension to crystals. Then, in Section 3 we exploit the symmetries of one- and two-body density functions to predict the scaling of one- and two-body entropies with the size of the crystal. The final Section 4 is reserved to concluding remarks.

2. Derivation of the Entropy MPCE

In this Section, we collect a number of well-established results on the entropy MPCE, with the only purpose of setting the language and notation for the rest of the paper. First, we recall the derivation of the entropy formula for a one-component system of classical particles in the canonical ensemble. Such an ensemble choice is by no means restrictive, since, as we show next, it is always possible to take advantage of the sum rules obeyed by the canonical correlation functions to arrange the entropy MPCE in an ensemble-invariant form. Then, in the following Section we present an application of the formalism to crystals.

The canonical partition function of a system of N classical particles of mass m at temperature T is $Z_N = Z_N^{\text{id}} Z_N^{\text{exc}}$, where the ideal and excess parts are given by

$$Z_N^{\text{id}} = \frac{1}{N!}\left(\frac{V}{\Lambda^3}\right)^N \quad \text{and} \quad Z_N^{\text{exc}} = \frac{1}{V^N}\int d^3 R_1 \cdots d^3 R_N \, e^{-\beta U(\mathbf{R}^N)}. \tag{1}$$

In Equation (1), V is the system volume, $\beta = 1/(k_B T)$, $\Lambda = h/\sqrt{2\pi m k_B T}$ is the thermal wavelength, and $U(\mathbf{R}^N)$ is an arbitrary potential energy. As the particles are identical, for each $n = 1, 2, \ldots, N$

the cumulative sum of all n-body terms in U is invariant under permutations of particle coordinates (we can also say that U is S_N-invariant, S_N being the symmetric group of the permutations on N symbols). The CE average of a function f of coordinates reads

$$\langle f(\mathbf{R}^N)\rangle \equiv \frac{1}{V^N}\int d^3R_1\cdots d^3R_N\, f(\mathbf{R}^N)\pi_{\text{can}}(\mathbf{R}^N) \quad \text{with} \quad \pi_{\text{can}}(\mathbf{R}^N) = \frac{e^{-\beta U(\mathbf{R}^N)}}{Z_N^{\text{exc}}}, \qquad (2)$$

where $\pi_{\text{can}}(\mathbf{R}^N)$ is the configurational part of the canonical density function. Finally, the excess entropy $S_N^{\text{exc}} \equiv S_N - S_N^{\text{id}}$ reads

$$\frac{S_N^{\text{exc}}}{k_B} = -\frac{1}{V^N}\int d^3R_1\cdots d^3R_N\,\pi_{\text{can}}(\mathbf{R}^N)\ln\pi_{\text{can}}(\mathbf{R}^N) = -\langle\ln\pi_{\text{can}}(\mathbf{R}^N)\rangle. \qquad (3)$$

We define a set of marginal density functions (MDFs) by

$$\begin{aligned} P^{(N)}(\mathbf{R}^N) &= \pi_{\text{can}}(\mathbf{R}^N); \\ P^{(n)}(\mathbf{R}^n) &= \frac{1}{V^{N-n}}\int d^3R_{n+1}\cdots d^3R_N\,\pi_{\text{can}}(\mathbf{R}^N) \quad (n=1,\ldots,N-1). \end{aligned} \qquad (4)$$

Owing to S_N-invariance of π_{can}, it makes no difference which vector radii are integrated out in Equation (4); hence, $P^{(n)}(\mathbf{r}^n)$ is S_n-invariant (for example, $P^{(2)}(\mathbf{r},\mathbf{r}') = P^{(2)}(\mathbf{r}',\mathbf{r})$). The following properties are obvious:

$$\frac{1}{V^n}\int d^3R_1\cdots d^3R_n\, P^{(n)}(\mathbf{R}^n) = 1 \quad \text{and} \quad \frac{1}{V}\int d^3R_{n+1}\, P^{(n+1)}(\mathbf{R}^{n+1}) = P^{(n)}(\mathbf{R}^n). \qquad (5)$$

Then, the n-body density functions (DFs), for $n = 1,\ldots,N$, can be expressed as

$$\rho^{(n)}(\mathbf{r}^n) \equiv \left\langle\sum_{i_1\ldots i_n}{}' \delta^3(\mathbf{R}_{i_1}-\mathbf{r}_1)\cdots\delta^3(\mathbf{R}_{i_n}-\mathbf{r}_n)\right\rangle = \frac{N!}{(N-n)!}\frac{P^{(n)}(\mathbf{r}^n)}{V^n}, \qquad (6)$$

where the sum in (6) is carried out over all n-tuples of distinct particles (for example, the sum for $n=2$ contains $N(N-1)$ terms). We note that $P^{(1)} = 1$ and $\rho^{(1)} \equiv N/V = \rho$ if no one-body term is present in U, i.e., if no external potential acts on the particles (then U is translationally invariant). $P^{(1)}(\mathbf{r})/V$ is the probability density of finding a particle in $\mathbf{r}$; hence, $\rho^{(1)}(\mathbf{r}) = NP^{(1)}(\mathbf{r})/V$ is the number density at $\mathbf{r}$. Similarly, $P^{(2)}(\mathbf{r},\mathbf{r}')/V^2$ is the probability density of finding one particle in $\mathbf{r}$ and another particle in $\mathbf{r}'$; hence, $\rho^{(2)}(\mathbf{r},\mathbf{r}') = N(N-1)P^{(2)}(\mathbf{r},\mathbf{r}')/V^2$ is the density of the number of particle pairs at $(\mathbf{r},\mathbf{r}')$. As $\mathbf{r}'$ increasingly departs from $\mathbf{r}$, the positions of two particles become less and less correlated, until $P^{(2)}(\mathbf{r},\mathbf{r}') = P^{(1)}(\mathbf{r})P^{(1)}(\mathbf{r}')$ at infinite distance. We stress that this cluster property holds in full generality, even for a broken-symmetry phase.

The n-body reduced density functions, for $n = 2,\ldots,N$, read

$$g^{(n)}(\mathbf{r}^n) \equiv \frac{\rho^{(n)}(\mathbf{r}^n)}{\rho^{(1)}(\mathbf{r}_1)\cdots\rho^{(1)}(\mathbf{r}_n)} = \left(1-\frac{1}{N}\right)\cdots\left(1-\frac{n-1}{N}\right)Q^{(n)}(\mathbf{r}^n)$$

$$\text{with} \quad Q^{(n)}(\mathbf{r}^n) = \frac{P^{(n)}(\mathbf{r}^n)}{P^{(1)}(\mathbf{r}_1)\cdots P^{(1)}(\mathbf{r}_n)}. \qquad (7)$$

These functions fulfill the property

$$\frac{1}{V}\int d^3R_{n+1}\, P^{(1)}(\mathbf{R}_{n+1})g^{(n+1)}(\mathbf{R}^{n+1}) = \left(1-\frac{n}{N}\right)g^{(n)}(\mathbf{R}^n), \qquad (8)$$

which also holds for $n=1$ if we define $g^{(1)} \equiv 1$. For a homogeneous fluid, $g^{(2)}(\mathbf{r},\mathbf{r}') = g(|\mathbf{r}-\mathbf{r}'|)$. From now on, we adopt the shorthand notation $P_{12\ldots n} = P^{(n)}(\mathbf{R}^n)$ and $Q_{12\ldots n} = Q^{(n)}(\mathbf{R}^n)$. Moreover, any integral of the kind $V^{-n} \int d^3R_1 \cdots d^3R_n (\cdots)$ is hereafter denoted as $\int(\cdots)$. For example, Equations (3) and (4) indicate that $S_N^{\text{exc}}/k_B = -\int P_{12\ldots N} \ln P_{12\ldots N}$.

To build up the CE expansion term by term, our strategy is to consider a progressively larger number of particles. For a one-particle system, the excess entropy in units of the Boltzmann constant is $S_1^{\text{exc}}/k_B = -\int P_1 \ln P_1$, leading to a first-order approximation to the excess entropy of a N-particle system in the form $S_N^{\text{exc}}/k_B \approx S_N^{(1)}/k_B \equiv -N\int P_1 \ln P_1$ (that is, each particle contributes to the entropy independently of the other particles). For a two-particle system, the excess entropy is $S_2^{(1)}$ plus a remainder $k_B R_2$, given by:

$$R_2 \equiv \frac{S_2^{\text{exc}} - S_2^{(1)}}{k_B} = -\int P_{12} \ln P_{12} + 2\int P_1 \ln P_1 = -\int P_{12} \ln Q_{12}. \tag{9}$$

Equation (9) suggests a second-order approximation for S_N^{exc}, where each distinct pair of particles contributes the same two-body residual term to the entropy:

$$\frac{S_N^{(2)}}{k_B} = -N\int P_1 \ln P_1 - \binom{N}{2}\int P_{12} \ln Q_{12}. \tag{10}$$

Notice that Equation (10) is exact for $N=2$, i.e., $S_2^{(2)} = S_2^{\text{exc}}$. Similarly, for a three-particle system the excess entropy is $S_3^{(2)}$ plus a remainder $k_B R_3$:

$$\begin{aligned} R_3 \equiv \frac{S_3^{\text{exc}} - S_3^{(2)}}{k_B} &= -\int P_{123} \ln P_{123} + 3\int P_1 \ln P_1 + \binom{3}{2}\int P_{12} \ln Q_{12} \\ &= -\int P_{123} \ln Q_{123} + \binom{3}{2}\int P_{12} \ln Q_{12}. \end{aligned} \tag{11}$$

Hence, a third-order approximation follows for S_N^{exc} in the form

$$\frac{S_N^{(3)}}{k_B} = -N\int P_1 \ln P_1 - \binom{N}{2}\int P_{12} \ln Q_{12} - \binom{N}{3}\left[\int P_{123} \ln Q_{123} - \binom{3}{2}\int P_{12} \ln Q_{12}\right]. \tag{12}$$

Again, $S_3^{(3)} = S_3^{\text{exc}}$. Equation (12) reproduces the first three terms in the rhs of Equation (5.9) of reference [8], and one may legitimately expect that the further terms in the entropy expansion are similarly obtained by arguing for $N = 4,5,\ldots$ like we did for $N = 1,2,3$ (see the proof in [8]).

The general entropy formula finally reads:

$$\begin{aligned} \frac{S_N^{\text{exc}}}{k_B} &= -\int P_{12\ldots N} \ln P_{12\ldots N} = -N\int P_1 \ln P_1 - \int P_{12\ldots N} \ln Q_{12\ldots N} \\ &= -N\int P_1 \ln P_1 - \sum_{n=2}^{N} \binom{N}{n} \sum_{a=2}^{n} (-1)^{n-a} \binom{n}{a} \int P_{1\ldots a} \ln Q_{1\ldots a}. \end{aligned} \tag{13}$$

This equation is trivially correct since, for any finite sequence $\{c_a\}$ of numbers,

$$c_N = \sum_{n=2}^{N} \binom{N}{n} \sum_{a=2}^{n} (-1)^{n-a} \binom{n}{a} c_a. \tag{14}$$

To prove (14), it is sufficient to observe that, for each fixed $k = 2, \ldots, N$, the coefficient of c_k in the above sum is

$$\sum_{n=k}^{N}(-1)^{n-k}\binom{N}{n}\binom{n}{k} = \sum_{n=0}^{N-k}(-1)^n\binom{N}{n+k}\binom{n+k}{k} = \binom{N}{k}\sum_{n=0}^{N-k}(-1)^n\binom{N-k}{n}$$
$$= \begin{cases} 0, & \text{for } 2 \leq k < N \\ 1, & \text{for } k = N. \end{cases} \quad (15)$$

A more compact entropy formula is

$$\frac{S_N^{\text{exc}}}{k_B} = -\sum_{n=1}^{N}\binom{N}{n}\sum_{a=1}^{n}(-1)^{n-a}\binom{n}{a}\int P_{1\ldots a}\ln P_{1\ldots a}, \quad (16)$$

which follows from

$$c_N = \sum_{n=1}^{N}\binom{N}{n}\sum_{a=1}^{n}(-1)^{n-a}\binom{n}{a}c_a. \quad (17)$$

The entropy expansion, (13) or (16), is only valid in the CE. Eliminating $Q_{1\ldots a}$ in favor of $g_{1\ldots a}$ by Equation (7), an overall constant comes out of the integral in Equation (13), namely,

$$\sum_{n=2}^{N}\binom{N}{n}\sum_{a=2}^{n}(-1)^{n-a}\binom{n}{a}\ln\frac{(N-1)\cdots(N-a+1)}{N^{a-1}}, \quad (18)$$

which, by Equation (15), equals $\ln(N!/N^N)$; this term exactly cancels an identical term present in the ideal-gas entropy. In the end, a modified entropy MPCE emerges:

$$\frac{S_N}{k_B} = N\left[\frac{3}{2} - \ln(\rho\Lambda^3)\right] - N\int P_1 \ln P_1 - \sum_{n=2}^{N}\binom{N}{n}\sum_{a=2}^{n}(-1)^{n-a}\binom{n}{a}\int P_{1\ldots a}\ln g_{1\ldots a}. \quad (19)$$

Notice that the first term in the rhs differs by N from the ideal-gas entropy expression in the thermodynamic limit. In order that Equation (19) conforms to the GCE entropy expansion, for each n a suitable fluctuation integral of value $-N/[n(n-1)]$ should be summed to (and subtracted from) the n-th term in the expansion. For example, using Equations (7) and (8) the second-order term in (13) can be rewritten as

$$-\binom{N}{2}\int\frac{d^3r_1 d^3r_2}{V^2}P_{12}\ln\frac{P_{12}}{P_1 P_2} = \binom{N}{2}\ln\frac{N-1}{N} - \binom{N}{2}\int\frac{d^3r_1 d^3r_2}{V^2}P_1 P_2 g_{12}\ln g_{12}$$
$$= \binom{N}{2}\ln\frac{N-1}{N} + \frac{N}{2} - \frac{1}{2}\rho^2\int d^3r_1 d^3r_2\, P_1 P_2 (g_{12}\ln g_{12} - g_{12} + 1). \quad (20)$$

Overall, the extra constants appearing in each term of the entropy formula (for example, the quantity $N/2$ in Equation (20)) add to N. By absorbing such a N in the first term of (19) we recover the ideal-gas entropy in the thermodynamic limit, and the CE expansion becomes formally identical to the grand-canonical MPCE [9].

In Appendix A we present another derivation of the entropy formula in the CE, which is closer in spirit to the one given by H. S. Green. In parallel, we show that the approximation obtained by truncating the MPCE at a given order can be derived from a modified $P_{12\ldots N}$ distribution, which is an explicit functional of the spatial correlation functions up to that order.

3. The First Few Terms in the Expansion of Crystal Entropy

The entropy expansion in the CE is formally identical for a fluid system and a crystal, since the origin of (13) is purely combinatorial. However, the DFs of the two phases are radically different: most notably, while $P_1 = 1$ and $\rho^{(1)} = \rho$ for a homogeneous fluid, the one-body density is spatially structured for a crystal—at least once the degeneracy due to translations and point-group operations has been lifted; we stress that $P_1 \neq 1$ only provided that a specific determination of the crystal is taken, since otherwise $P_1 = 1$ also in the "delocalized" crystalline phase. In practice, in order to fix a crystal in space we should imagine to apply a suitable symmetry-breaking external potential, whose strength is sent to zero after statistical averages have been carried out (in line with Bogoliubov's advice to interpret statistical averages of broken-symmetry phases as quasiaverages [28], which amounts to sending the strength of the symmetry-breaking potential to zero only after the thermodynamic limit has been taken). A way to accomplish this task is to constrain the position of just one particle. When periodic boundary conditions are applied, keeping one particle fixed will be enough to break the continuous symmetries of free space. As N grows, the effect of the external potential becomes weaker and weaker, since it does not scale with the size of the system.

3.1. One-Body Entropy

A reasonable form of one-body density for a three-dimensional Bravais crystal without defects is the Tarazona ansatz [29]:

$$\rho^{(1)}(\mathbf{r}) = \left(\frac{\alpha}{\pi}\right)^{3/2} \sum_{\mathbf{R}} e^{-\alpha(\mathbf{r}-\mathbf{R})^2} = \rho \sum_{\mathbf{G}} e^{-\frac{G^2}{4\alpha}} e^{i\mathbf{G}\cdot\mathbf{r}}, \tag{21}$$

where $\alpha > 0$ is a temperature-dependent parameter, the $\mathbf{R}$'s are direct-lattice vectors, and the $\mathbf{G}$'s are reciprocal-lattice vectors (recall that $\mathbf{G} \cdot \mathbf{R} = 2\pi m$ with $m \in \mathbb{Z}$ and $\int_V d^3r \exp\{i(\mathbf{G}+\mathbf{G}')\cdot\mathbf{r}\} = V\delta_{\mathbf{G}',-\mathbf{G}}$). Equation (21) is a rather generic form of crystal density, which recently we have also applied in a different context [30]. More generally, the one-body density appropriate to a perfect crystal must obey $\rho^{(1)}(\mathbf{r}+\mathbf{R}) = \rho^{(1)}(\mathbf{r})$ for all $\mathbf{R}$, and is thus necessarily of the form

$$\rho^{(1)}(\mathbf{r}) = \sum_{\mathbf{G}} \tilde{u}_{\mathbf{G}} e^{i\mathbf{G}\cdot\mathbf{r}} \quad \text{with} \quad \tilde{u}_{\mathbf{G}}^* = \tilde{u}_{-\mathbf{G}}. \tag{22}$$

Since $\int_V d^3r \rho^{(1)}(\mathbf{r}) = N$, it soon follows $\tilde{u}_0 = \rho$. Calling $\mathcal{C}$ a primitive cell and v_0 its volume, $\tilde{u}_{\mathbf{G}} = v_0^{-1} \int_{\mathcal{C}} d^3r \rho^{(1)}(\mathbf{r}) \exp\{-i\mathbf{G}\cdot\mathbf{r}\} \to 0$ as $G \to \infty$ (by the Riemann–Lebesgue lemma). In real space, a legitimate $\rho^{(1)}(\mathbf{r})$ function is $\sum_{\mathbf{R}} \phi(\mathbf{r}-\mathbf{R})$ with $\int d^3r \phi(\mathbf{r}) = 1$ (integration bounds are left unspecified when the integral is over a macroscopic V). In the zero-temperature/infinite-density limit, particles sit at the lattice sites and the one-body density then becomes

$$\rho^{(1)}(\mathbf{r}) = \sum_{\mathbf{R}} \delta^3(\mathbf{r}-\mathbf{R}). \tag{23}$$

Equation (23) is also recovered from Equation (21) in the $\alpha \to \infty$ limit.

For a crystalline solid, the one-body entropy, that is, the first term in the expansion of excess entropy, is (in units of k_B):

$$S_1 \equiv -\frac{N}{V} \int d^3r_1 P_1 \ln P_1 = -\int d^3r_1 \rho^{(1)}(\mathbf{r}_1) \ln \frac{\rho^{(1)}(\mathbf{r}_1)}{\rho}. \tag{24}$$

One may wonder whether the integral in (24) is $\mathcal{O}(N)$ in the infinite-size limit. The answer is affirmative, and a simple argument goes as follows. Let $\rho^{(1)}(\mathbf{r})$ be $\sum_{\mathbf{R}} \phi(\mathbf{r}-\mathbf{R})$; if $\phi(\mathbf{r})$ is strongly localized near $\mathbf{r} = 0$, then $\rho^{(1)}(\mathbf{r}) \simeq \phi(\mathbf{r})$ in the cell around $\mathbf{R} = 0$ and $S_1 \simeq -N \int_{\mathcal{C}} d^3r \phi(\mathbf{r}) \ln\{\phi(\mathbf{r})/\rho\} = \mathcal{O}(N)$. Actually, we can provide a rigorous proof that S_1 is negative-semidefinite and its absolute value does not grow faster than N. Using $\ln x \leq x - 1$ for $x > 0$ and $x \ln x \geq x - 1$ for any $x \geq 0$, we obtain

$$0 = \rho \int d^3r_1 \left(\frac{\rho_1}{\rho} - 1\right) \leq \int d^3r_1 \rho_1 \ln \frac{\rho_1}{\rho} \leq \int d^3r_1 \rho_1 \left(\frac{\rho_1}{\rho} - 1\right) = \rho^{-1} \int d^3r_1 \rho_1^2 - N. \tag{25}$$

To estimate $\int d^3r_1 \rho_1^2$ we employ the one-body density in (21), which is sufficiently generic for our purposes:

$$\rho^{-1} \int d^3 r \left(\rho^{(1)}(\mathbf{r})\right)^2 = \rho \sum_{\mathbf{G},\mathbf{G}'} e^{-\left(\frac{G^2}{4\alpha} + \frac{G'^2}{4\alpha}\right)} \int d^3 r\, e^{i(\mathbf{G}+\mathbf{G}')\cdot \mathbf{r}} = N \sum_{\mathbf{G}} e^{-\frac{G^2}{2\alpha}} \quad (26)$$

(the above result is nothing but Parseval's theorem as applied to (21)). The sum in the rhs of Equation (26) is the three-dimensional analog of a Jacobi theta function (see, e.g., [31]), whose value is $\mathcal{O}(1)$ for $\alpha > 0$. Therefore, it follows from Equations (25) and (26) that the one-body entropy is at most $\mathcal{O}(N)$.

3.2. Two-Body Entropy

We now move to the problem of evaluating the two-body entropy S_2 for a crystal. For a homogeneous fluid, S_2 is an extensive quantity which, in k_B units, is equal to

$$\text{fluid}: \quad S_2 = -2\pi\rho N \int_0^\infty dr\, r^2 \left(g(r)\ln g(r) - g(r) + 1\right). \quad (27)$$

For a crystal, we have from Equation (20) that

$$S_2 = -\frac{1}{2}\rho^2 \int d^3r_1 d^3r_2\, P_1 P_2 \left(g_{12}\ln g_{12} - g_{12} + 1\right). \quad (28)$$

As $x \ln x \geq x - 1$ for $x > 0$, S_2 is usually negative and zero exclusively for $g_{12} = 1$. In terms of density functions, S_2 is written as

$$S_2 = -\frac{1}{2}\int d^3r_1 d^3r_2 \left(\rho^{(2)}(\mathbf{r}_1,\mathbf{r}_2)\ln \frac{\rho^{(2)}(\mathbf{r}_1,\mathbf{r}_2)}{\rho^{(1)}(\mathbf{r}_1)\rho^{(1)}(\mathbf{r}_2)} - \rho^{(2)}(\mathbf{r}_1,\mathbf{r}_2) + \rho^{(1)}(\mathbf{r}_1)\rho^{(1)}(\mathbf{r}_2)\right). \quad (29)$$

We show below that Equation (29) can be expressed as a radial integral, i.e., in a way similar to the two-body entropy for a fluid.

We can assign a radial structure to crystals by appealing to a couple of functions introduced in [32], namely

$$\rho^2 \widetilde{g}(r) = \int \frac{d^3r_1}{V} \int \frac{d^2\Omega}{4\pi} \rho^{(2)}(\mathbf{r}_1, \mathbf{r}_1 + \mathbf{r}) \quad (30)$$

and

$$\rho^2 \widetilde{g}_0(r) = \int \frac{d^3r_1}{V} \int \frac{d^2\Omega}{4\pi} \rho^{(1)}(\mathbf{r}_1)\rho^{(1)}(\mathbf{r}_1 + \mathbf{r}), \quad (31)$$

where the inner integrals are over the direction of $\mathbf{r}$. For a homogeneous fluid, $\widetilde{g}(r) = g(r)$ and $\widetilde{g}_0(r) = 1$. The authors of reference [32] have sketched the profile of $\widetilde{g}(r)$ and $\widetilde{g}_0(r)$ for a crystal; both functions show narrow peaks at neighbor positions in the lattice, with an extra peak at zero distance for $\widetilde{g}_0(r)$, and the oscillations persist till large distances. The following sum rules hold (cf. Equation (8) for $n = 1$):

$$4\pi \int dr\, r^2 \rho \widetilde{g}(r) = \frac{1}{\rho} 4\pi \int \frac{d^3r_1}{V} \rho^{(1)}(\mathbf{r}_1) \underbrace{\int \frac{d^3r_2}{4\pi} \frac{\rho^{(2)}(\mathbf{r}_1,\mathbf{r}_2)}{\rho^{(1)}(\mathbf{r}_1)}}_{\frac{N-1}{4\pi}} = N - 1 \quad (32)$$

and

$$4\pi \int dr\, r^2 \rho \widetilde{g}_0(r) = \frac{1}{\rho} 4\pi \int \frac{d^3r_1}{V} \rho^{(1)}(\mathbf{r}_1) \underbrace{\int \frac{d^3r_2}{4\pi} \rho^{(1)}(\mathbf{r}_2)}_{\frac{N}{4\pi}} = N. \quad (33)$$

When the latter two formulae are rewritten as

$$4\pi\rho \int dr\, r^2 (\tilde{g}(r) - 1) = -1 \quad \text{and} \quad 4\pi\rho \int dr\, r^2 (\tilde{g}_0(r) - 1) = 0, \tag{34}$$

it becomes apparent that both $\tilde{g}(r)$ and $\tilde{g}_0(r)$ decay to 1 at infinity. Similarly, we define:

$$\rho^2 \tilde{h}(r) = \int \frac{d^3 r_1}{V} \int \frac{d^2\Omega}{4\pi} \rho^{(2)}(\mathbf{r}_1, \mathbf{r}_1 + \mathbf{r}) \ln \frac{\rho^{(2)}(\mathbf{r}_1, \mathbf{r}_1 + \mathbf{r})}{\rho^{(1)}(\mathbf{r}_1)\rho^{(1)}(\mathbf{r}_1 + \mathbf{r})}, \tag{35}$$

which obviously vanishes at infinity. While $\tilde{h}(r) = g(r) \ln g(r)$ for a homogeneous fluid, we expect that $\tilde{h}(r) \neq \tilde{g}(r) \ln \tilde{g}(r)$ in the crystal. Putting Equations (30)–(35) together, we arrive at

$$\text{crystal}: \quad S_2 = -2\pi\rho N \int_0^\infty dr\, r^2 \left(\tilde{h}(r) - \tilde{g}(r) + \tilde{g}_0(r) \right) = -2\pi\rho N \int_0^\infty dr\, r^2 \tilde{h}(r) - \frac{N}{2}. \tag{36}$$

Even though the integrand vanishes at infinity, $S_2 = \mathcal{O}(N)$ only if the envelope of $\tilde{h}(r)$ decays faster than r^{-3} (r^{-2} in two dimensions). A slower decay may be sufficient if S_2 is computed through the first integral in (36). For a spherically-symmetric interaction potential, the excess energy (i.e., the canonical average of the total potential energy U) can also be written as a radial integral:

$$\begin{aligned}
\langle U \rangle &= \frac{1}{2} \int d^3 r_1 d^3 r_2\, \rho^{(2)}(\mathbf{r}_1, \mathbf{r}_2) u(|\mathbf{r}_2 - \mathbf{r}_1|) \\
&= \frac{1}{2} \int_0^\infty dr\, r^2 u(r) \int d^3 r_1 \int d^2\Omega\, \rho^{(2)}(\mathbf{r}_1, \mathbf{r}_1 + \mathbf{r}) = 2\pi\rho N \int_0^\infty dr\, r^2 u(r) \tilde{g}(r).
\end{aligned} \tag{37}$$

For the one-body density in (21), $\tilde{g}_0(r)$ can be obtained in closed form. First we have:

$$\int \frac{d^2\Omega}{4\pi} \rho^{(1)}(\mathbf{r}_1 + \mathbf{r}) = \rho \sum_{\mathbf{G}} e^{-\frac{G^2}{4\alpha}} e^{i\mathbf{G}\cdot\mathbf{r}_1} \frac{\sin(Gr)}{Gr}. \tag{38}$$

Then, multiplying by $\rho^{(1)}(\mathbf{r}_1) = \rho \sum_{\mathbf{G}'} e^{-\frac{G'^2}{4\alpha}} e^{i\mathbf{G}'\cdot\mathbf{r}_1}$ and finally integrating over $\mathbf{r}_1$ we arrive at

$$\tilde{g}_0(r) = 1 + \sum_{\mathbf{G}\neq 0} e^{-\frac{G^2}{2\alpha}} \frac{\sin(Gr)}{Gr}. \tag{39}$$

We see that the large-distance decay of $\tilde{g}_0(r)$ is usually slow, and the same will occur for $\tilde{g}(r)$ since $\tilde{g}(r) \simeq \tilde{g}_0(r)$ for large r. In two dimensions, the one-body density and $\tilde{g}_0$ functions respectively read:

$$\rho^{(1)}(\mathbf{r}) = \frac{\alpha}{\pi} \sum_{\mathbf{R}} e^{-\alpha(\mathbf{r}-\mathbf{R})^2} = \rho \sum_{\mathbf{G}} e^{-\frac{G^2}{4\alpha}} e^{i\mathbf{G}\cdot\mathbf{r}} \quad \text{and} \quad \tilde{g}_0(r) = 1 + \sum_{\mathbf{G}\neq 0} e^{-\frac{G^2}{2\alpha}} J_0(Gr), \tag{40}$$

where J_0 is a Bessel function of the first kind. Since the envelope of J_0 maxima decays as $r^{-1/2}$ at infinity, we see that the asymptotic vanishing of $\tilde{g}_0$ is slower in two dimensions than in three.

Equation (39) has a definite limit for $\alpha \to \infty$, corresponding to zero temperature. Indeed, using Poisson summation formula and the expression of Dirac's delta in spherical coordinates, we obtain:

$$\begin{aligned}
\rho \tilde{g}_0(r) &= \rho \left(1 + \sum_{\mathbf{G}\neq 0} \frac{\sin(Gr)}{Gr} \right) = \rho \int \frac{d^2\Omega}{4\pi} \sum_{\mathbf{G}} e^{i\mathbf{G}\cdot\mathbf{r}} = \delta^3(\mathbf{r}) + \int \frac{d^2\Omega}{4\pi} \sum_{\mathbf{R}\neq 0} \delta^3(\mathbf{r} - \mathbf{R}) \\
&= \delta^3(\mathbf{r}) + \sum_{\mathbf{R}\neq 0} \frac{1}{4\pi} \int_0^{2\pi} d\phi \int_0^\pi d\theta \sin\theta \frac{1}{r^2 \sin\theta} \delta(r - R)\delta(\theta - \theta_\mathbf{R})\delta(\phi - \phi_\mathbf{R}) \\
&= \delta^3(\mathbf{r}) + \sum_{\mathbf{R}\neq 0} \frac{1}{4\pi R^2} \delta(r - R).
\end{aligned} \tag{41}$$

Hence, $\widetilde{g}_0(r)$ reduces to a sum of delta functions centered at lattice distances (including the origin). The latter result is actually general. Inserting Equation (23) in (31), we obtain:

$$\begin{aligned}
\rho \widetilde{g}_0(r) &= \frac{1}{\rho} \int \frac{d^3 r_1}{V} \int \frac{d^2 \Omega}{4\pi} \sum_{\mathbf{R}} \delta^3(\mathbf{r}_1 - \mathbf{R}) \sum_{\mathbf{R}'} \delta^3(\mathbf{r}_1 + \mathbf{r} - \mathbf{R}') \\
&= \frac{1}{\rho} \int \frac{d^3 r_1}{V} \int \frac{d^2 \Omega}{4\pi} \left\{ \sum_{\mathbf{R}} \delta^3(\mathbf{r}_1 - \mathbf{R}) \delta^3(\mathbf{r}_1 + \mathbf{r} - \mathbf{R}) + \sum_{\mathbf{R} \neq \mathbf{R}'} \delta^3(\mathbf{r}_1 - \mathbf{R}) \delta^3(\mathbf{r}_1 + \mathbf{r} - \mathbf{R}') \right\} \\
&= \frac{1}{\rho} \int \frac{d^3 r_1}{V} \int \frac{d^2 \Omega}{4\pi} \left\{ \sum_{\mathbf{R}} \delta^3(\mathbf{r}_1 - \mathbf{R}) \delta^3(\mathbf{r}) + \sum_{\mathbf{R} \neq \mathbf{R}'} \delta^3(\mathbf{r}_1 - \mathbf{R}) \delta^3(\mathbf{r}_1 + \mathbf{r} - \mathbf{R}') \right\} \\
&= \frac{1}{\rho} \sum_{\mathbf{R}} \delta^3(\mathbf{r}) \int \frac{d^3 r_1}{V} \delta^3(\mathbf{r}_1 - \mathbf{R}) + \frac{1}{\rho} \sum_{\mathbf{R} \neq \mathbf{R}'} \int \frac{d^3 r_1}{V} \delta^3(\mathbf{r}_1 - \mathbf{R}) \frac{\delta(r - |\mathbf{r}_1 - \mathbf{R}'|)}{4\pi |\mathbf{r}_1 - \mathbf{R}'|^2} \\
&= \delta^3(\mathbf{r}) + \sum_{\mathbf{R} \neq 0} \frac{1}{4\pi R^2} \delta(r - R),
\end{aligned} \quad (42)$$

q.e.d. At zero temperature, $\rho \widetilde{g}(r)$ is given by the same sum of delta-function terms as in (42), but for the first term, $\delta^3(\mathbf{r})$, which is missing—see Equation (92) below.

We add a final comment on possible alternative formulations of $\widetilde{g}(r)$ for a crystal. One choice is to replace (30) with

$$\text{option B:} \quad \rho \widetilde{g}(r) = \int \frac{d^3 r_1}{V} \int \frac{d^2 \Omega}{4\pi} \frac{\rho^{(2)}(\mathbf{r}_1, \mathbf{r}_1 + \mathbf{r})}{\rho^{(1)}(\mathbf{r}_1)}. \quad (43)$$

Apparently, this is a good definition since (see Equation (8))

$$4\pi \int dr\, r^2 \rho \widetilde{g}(r) = 4\pi \underbrace{\int \frac{d^3 r_1}{V}}_{1} \underbrace{\int \frac{d^3 r_2}{4\pi} \frac{\rho^{(2)}(\mathbf{r}_1, \mathbf{r}_2)}{\rho^{(1)}(\mathbf{r}_1)}}_{\frac{N-1}{4\pi}} = N - 1. \quad (44)$$

However, with this $\widetilde{g}(r)$ we cannot write S_2 as a radial integral—hence, option B is discarded altogether. Another possibility is

$$\text{option C:} \quad \widetilde{g}(r) = \int \frac{d^3 r_1}{V} \int \frac{d^2 \Omega}{4\pi} \frac{\rho^{(2)}(\mathbf{r}_1, \mathbf{r}_1 + \mathbf{r})}{\rho^{(1)}(\mathbf{r}_1) \rho^{(1)}(\mathbf{r}_1 + \mathbf{r})}, \quad (45)$$

but this option is useless too, since

$$4\pi \int dr\, r^2 \rho \widetilde{g}(r) = \rho \int d^3 r_1 \frac{1}{V} \int d^3 r_2\, g^{(2)}(\mathbf{r}_1, \mathbf{r}_2) = ? \quad (46)$$

(observe that the inner integral is different from the one appearing in Equation (8)).

3.3. Symmetries of the Two-Body Density

A general property of the two-body density for a crystal is the CE sum rule

$$\int d^3 r_2\, \rho^{(2)}(\mathbf{r}_1, \mathbf{r}_2) = (N-1) \rho^{(1)}(\mathbf{r}_1). \quad (47)$$

Other constraints follow from the translational symmetry of local crystal properties. As for the one-body density, fulfilling $\rho^{(1)}(\mathbf{r}_1 + \mathbf{R}) = \rho^{(1)}(\mathbf{r}_1)$ for every $\mathbf{R}$, we must have that

$$\rho^{(2)}(\mathbf{r}_1, \mathbf{r}_2) = \rho^{(2)}(\mathbf{r}_1 + \mathbf{R}, \mathbf{r}_2 + \mathbf{R}), \quad (48)$$

in turn implying

$$g^{(2)}(\mathbf{r}_1, \mathbf{r}_2) = g^{(2)}(\mathbf{r}_1 + \mathbf{R}, \mathbf{r}_2 + \mathbf{R}). \quad (49)$$

Now observe [33] that (i) any function of $\mathbf{r}_1$ and $\mathbf{r}_2$ can also be viewed as a function of $(\mathbf{r}_1 + \mathbf{r}_2)/2$ and $\mathbf{r}_2 - \mathbf{r}_1$; (ii) under a $\mathbf{R}$-translation, only the former variable is affected, not the relative separation. Hence, the most general function consistent with (49) is:

$$g^{(2)}(\mathbf{r}_1, \mathbf{r}_2) = \sum_{\mathbf{G}} \widetilde{v}_{\mathbf{G}}(\mathbf{r}_2 - \mathbf{r}_1) e^{i\mathbf{G} \cdot \frac{\mathbf{r}_1 + \mathbf{r}_2}{2}}, \tag{50}$$

where

$$g^{(2)}(\mathbf{r}_1, \mathbf{r}_2) \in \mathbb{R} \implies \widetilde{v}_{\mathbf{G}}^*(\mathbf{r}_2 - \mathbf{r}_1) = \widetilde{v}_{-\mathbf{G}}(\mathbf{r}_2 - \mathbf{r}_1) \tag{51}$$

and

$$g^{(2)}(\mathbf{r}_1, \mathbf{r}_2) = g^{(2)}(\mathbf{r}_2, \mathbf{r}_1) \implies \widetilde{v}_{\mathbf{G}}(\mathbf{r}_2 - \mathbf{r}_1) = \widetilde{v}_{\mathbf{G}}(\mathbf{r}_1 - \mathbf{r}_2). \tag{52}$$

In order that $\lim_{r \to \infty} g^{(2)}(\mathbf{r}_1, \mathbf{r}_1 + \mathbf{r}) = 1$ it is sufficient that

$$\lim_{r \to \infty} \widetilde{v}_0(\mathbf{r}) = 1 \quad \text{and} \quad \lim_{r \to \infty} \widetilde{v}_{\mathbf{G}}(\mathbf{r}) = 0 \text{ for } \mathbf{G} \neq 0. \tag{53}$$

We may reasonably expect that the most relevant term in the expansion (50) is indeed the $\mathbf{G} = 0$ one (also notice that $\widetilde{v}_{\mathbf{G}} \to 0$ as $G \to \infty$ by the Riemann–Lebesgue lemma).

Equation (50) is still insufficient to establish the scaling of two-body entropy with the size of the crystal. Some general results can be obtained under the (strong) assumption that $\widetilde{v}_{\mathbf{G}}(\mathbf{r}) = 0$ for any $\mathbf{G} \neq 0$. If we change the notation from $\widetilde{v}_0$ to $\mathcal{G}(\mathbf{r}) \equiv 1 + \mathcal{H}(\mathbf{r})$ (which, by Equations (51) and (52), is a real and even function), then a necessary condition for $\mathcal{H}$ is:

$$\int d^3 r_2 \, \rho^{(1)}(\mathbf{r}_2) \mathcal{H}(\mathbf{r}_2 - \mathbf{r}_1) = -1 \quad \text{for any } \mathbf{r}_1 \text{ where } \rho^{(1)}(\mathbf{r}_1) \neq 0. \tag{54}$$

The rationale behind Equation (54) is particularly transparent near $T = 0$, where the peaks of the one-body density are extremely narrow. As argued below (see Equation (67) ff.), $\mathcal{H}$ as a function of $\mathbf{r}_2$ is roughly -1 in the primitive cell $\mathcal{C}$ centered in $\mathbf{r}_1 \approx \mathbf{R}_1$, $\mathbf{R}_1$ denoting the only lattice site contained in $\mathcal{C}$ and roughly zero outside $\mathcal{C}$. Since the integral of $\rho^{(1)}$ over $\mathcal{C}$ equals 1, Equation (54) will immediately follow.

Now writing $\mathcal{H}(\mathbf{r})$ as a Fourier integral,

$$\mathcal{H}(\mathbf{r}) = \int \frac{d^3 k}{(2\pi)^3} \widetilde{\mathcal{H}}(\mathbf{k}) e^{i\mathbf{k} \cdot \mathbf{r}}, \tag{55}$$

and using (21) as one-body density, Equation (54) yields

$$\rho \sum_{\mathbf{G}} e^{-\frac{G^2}{4\alpha}} \widetilde{\mathcal{H}}(\mathbf{G}) e^{-i\mathbf{G} \cdot \mathbf{r}_1} = -1, \tag{56}$$

which can only hold for arbitrary $\mathbf{r}_1$ if

$$\widetilde{\mathcal{H}}(\mathbf{G}) = -\frac{1}{\rho} \delta_{\mathbf{G},0}. \tag{57}$$

Next, from Equation (30) we obtain:

$$\rho^2 \widetilde{g}(r) = \rho^2 \widetilde{g}_0(r) + \int \frac{d^3 r_1}{V} \rho^{(1)}(\mathbf{r}_1) \int \frac{d^2 \Omega}{4\pi} \rho^{(1)}(\mathbf{r}_1 + \mathbf{r}) \mathcal{H}(\mathbf{r}). \tag{58}$$

For the one-body density in (21), the inner integral becomes:

$$\int \frac{d^2 \Omega}{4\pi} \rho^{(1)}(\mathbf{r}_1 + \mathbf{r}) \mathcal{H}(\mathbf{r}) = \rho \sum_{\mathbf{G}} e^{-\frac{G^2}{4\alpha}} I_{\mathbf{G}}(r) e^{-i\mathbf{G} \cdot \mathbf{r}_1} \tag{59}$$

with
$$I_{\mathbf{G}}(r) = \int \frac{d^2\Omega}{4\pi} \mathcal{H}(\mathbf{r}) e^{-i\mathbf{G}\cdot\mathbf{r}} = \int \frac{d^3k}{(2\pi)^3} \widetilde{\mathcal{H}}(\mathbf{k}) \frac{\sin(|\mathbf{k}-\mathbf{G}|r)}{|\mathbf{k}-\mathbf{G}|r}. \tag{60}$$

It is evident that $I_{\mathbf{G}}(r)$ vanishes at infinity. Upon inserting (59) in (58), we finally obtain:

$$\widetilde{g}(r) = \widetilde{g}_0(r) + \sum_{\mathbf{G}} e^{-\frac{G^2}{2\alpha}} I_{\mathbf{G}}(r). \tag{61}$$

As r increases, the second term gradually vanishes and the large-distance oscillations of $\widetilde{g}(r)$ then exactly match those of $\widetilde{g}_0(r)$. As a countercheck, let us compute the integral of $\rho\widetilde{g}(r) - \rho\widetilde{g}_0(r)$ over the macroscopic system volume (which, by Equations (32) and (33), should be -1):

$$\begin{aligned}
4\pi \int dr\, r^2 \rho(\widetilde{g}(r) - \widetilde{g}_0(r)) &= \rho \sum_{\mathbf{G}} e^{-\frac{G^2}{2\alpha}} \cdot 4\pi \int dr\, r^2 I_{\mathbf{G}}(r) \\
&= \rho \sum_{\mathbf{G}} e^{-\frac{G^2}{2\alpha}} \int \frac{d^3k}{(2\pi)^3} \widetilde{\mathcal{H}}(\mathbf{k}) \cdot 4\pi \int dr\, r^2 \frac{\sin(|\mathbf{k}-\mathbf{G}|r)}{|\mathbf{k}-\mathbf{G}|r} \\
&= \rho \sum_{\mathbf{G}} e^{-\frac{G^2}{2\alpha}} \int \frac{d^3k}{(2\pi)^3} \widetilde{\mathcal{H}}(\mathbf{k}) \underbrace{\int d^3r\, e^{i(\mathbf{k}-\mathbf{G})\cdot\mathbf{r}}}_{(2\pi)^3 \delta^3(\mathbf{k}-\mathbf{G})} \\
&= \rho \sum_{\mathbf{G}} e^{-\frac{G^2}{2\alpha}} \underbrace{\widetilde{\mathcal{H}}(\mathbf{G})}_{-(1/\rho)\delta_{\mathbf{G},0}} = -1.
\end{aligned} \tag{62}$$

Under the assumption that
$$\rho^{(2)}(\mathbf{r}_1, \mathbf{r}_1 + \mathbf{r}) = \rho^{(1)}(\mathbf{r}_1) \rho^{(1)}(\mathbf{r}_1 + \mathbf{r}) \mathcal{G}(\mathbf{r}), \tag{63}$$

the entropy expansion for a crystal reads

$$\begin{aligned}
\frac{S}{k_B} =\ & -\int d^3r_1 \rho^{(1)}(\mathbf{r}_1) \ln \frac{\rho^{(1)}(\mathbf{r}_1)}{\rho} \\
& -\frac{1}{2} \int d^3r_1 \rho^{(1)}(\mathbf{r}_1) \int d^3r\, \rho^{(1)}(\mathbf{r}_1 + \mathbf{r}) \left[\mathcal{G}(\mathbf{r}) \ln \mathcal{G}(\mathbf{r}) - \mathcal{G}(\mathbf{r}) + 1\right] + \ldots
\end{aligned} \tag{64}$$

Providing that it vanishes sufficiently rapidly at infinity, the function

$$\mathcal{K}(\mathbf{r}) = \mathcal{G}(\mathbf{r}) \ln \mathcal{G}(\mathbf{r}) - \mathcal{G}(\mathbf{r}) + 1 \tag{65}$$

can be written as a Fourier integral, and using (21) as one-body density, the two-body entropy becomes

$$S_2 = -\frac{1}{2}\rho^2 \sum_{\mathbf{G},\mathbf{G}'} e^{-\frac{G^2+G'^2}{4\alpha}} \underbrace{\int d^3r_1\, e^{i(\mathbf{G}+\mathbf{G}')\cdot\mathbf{r}_1}}_{V\delta_{\mathbf{G}',-\mathbf{G}}} \int d^3r\, \mathcal{K}(\mathbf{r}) e^{i\mathbf{G}'\cdot\mathbf{r}} = -\frac{1}{2} N\rho \sum_{\mathbf{G}} e^{-\frac{G^2}{2\alpha}} \widetilde{\mathcal{K}}(\mathbf{G}), \tag{66}$$

which is clearly $\mathcal{O}(N)$.

3.4. Two-Body Density at $T = 0$

In the zero-temperature limit, particles will be sitting at lattice sites, and the two-body density then becomes (see Equation (23)):

$$\rho^{(2)}(\mathbf{r}_1, \mathbf{r}_2) = \sideset{}{'}\sum_{\mathbf{R},\mathbf{R}'} \delta^3(\mathbf{r}_1 - \mathbf{R}) \delta^3(\mathbf{r}_2 - \mathbf{R}') = \rho^{(1)}(\mathbf{r}_1) \rho^{(1)}(\mathbf{r}_2) \left(1 - \mathbf{1}_{\mathcal{C}}(\mathbf{r}_2 - \mathbf{r}_1)\right), \tag{67}$$

which is of the form (63). In Equation (67), $\mathbf{1}_\mathcal{C}(\mathbf{r})$ is the indicator function of a Wigner–Seitz cell $\mathcal{C}$ centered at the origin (i.e., $\mathbf{1}_\mathcal{C}(\mathbf{r}) = 1$ if $\mathbf{r} \in \mathcal{C}$ and $\mathbf{1}_\mathcal{C}(\mathbf{r}) = 0$ otherwise). While the factor $\rho^{(1)}(\mathbf{r}_1)\rho^{(1)}(\mathbf{r}_2)$ forces particles to be located at lattice sites, the only role of the $\mathcal{G}$ in (67) is to prevent the possibility of double site occupancy. However, a $\mathcal{G}$ function with this property is not unique; the one provided in (67) has the advantage of exactly complying with condition (57) (see below). Equation (67) indicates that the pair-correlation structure of a low-temperature solid is very different from the structure of a dense fluid close to freezing.

For
$$\mathcal{H}(\mathbf{r}) = -\mathbf{1}_\mathcal{C}(\mathbf{r}) = \begin{cases} -1, & \text{for } \mathbf{r} \in \mathcal{C} \\ 0, & \text{otherwise} \end{cases} \tag{68}$$

the Fourier transform reads:
$$\widetilde{\mathcal{H}}(\mathbf{k}) = \int d^3r\, \mathcal{H}(\mathbf{r}) e^{-i\mathbf{k}\cdot\mathbf{r}} = -\int_\mathcal{C} d^3r\, e^{-i\mathbf{k}\cdot\mathbf{r}}. \tag{69}$$

Now observe that $f(\mathbf{r}) = 1$ is trivially periodic, and can thus be expanded in plane waves as $1 = \sum_\mathbf{G} \widetilde{f}_\mathbf{G} e^{i\mathbf{G}\cdot\mathbf{r}}$, with $\widetilde{f}_\mathbf{G} = \delta_{\mathbf{G},0}$. On the other hand,
$$\widetilde{f}_\mathbf{G} = \frac{1}{v_0}\int_\mathcal{C} d^3r\, f(\mathbf{r}) e^{-i\mathbf{G}\cdot\mathbf{r}} = \rho \int_\mathcal{C} d^3r\, e^{-i\mathbf{G}\cdot\mathbf{r}}. \tag{70}$$

Comparing Equations (69) and (70), we conclude that
$$\widetilde{\mathcal{H}}(\mathbf{G}) = -\frac{1}{\rho}\delta_{\mathbf{G},0}. \tag{71}$$

For $\mathcal{H}(\mathbf{r}) = -\mathbf{1}_\mathcal{C}(\mathbf{r})$ the function $I_\mathbf{G}(r)$ at Equation (60) equals $-\sin(Gr)/(Gr)$ for $r < r_m$ and 0 for $r > r_M$, where r_m (r_M) is the radius of the largest (smallest) sphere inscribed in (circumscribed to) $\mathcal{C}$. It then follows from Equation (61) that $\widetilde{g}(r) = 0$ for $r < r_m$, while $\widetilde{g}(r) = \widetilde{g}_0(r)$ for $r > r_M$ (for a triangular crystal with spacing a we have $r_m = a/2$ and $r_M = a/\sqrt{3}$, both comprised between the first, 0, and the second, a, lattice distance). $T = 0$, where $\widetilde{g}_0(r)$ consists of infinitely narrow peaks centered at lattice distances; this implies that $\widetilde{g}(r) = \widetilde{g}_0(r)$ everywhere but at the origin, where $\widetilde{g}(r) = 0$ while $\widetilde{g}_0(r)$ is non-zero.

3.5. Scaling of Two-Body Entropy with N

We henceforth discuss in fully general terms how the two-body entropy scales with N for a crystal, avoiding to make any simplifying hypothesis on the structure of $g^{(2)}(\mathbf{r}_1, \mathbf{r}_2)$. Using an obvious short-hand notation, the two-body entropy reads
$$S_2 = -\frac{1}{2}\int d1\, d2\left(\rho_{12}\ln\frac{\rho_{12}}{\rho_1\rho_2} - \rho_{12} + \rho_1\rho_2\right) = -\frac{1}{2}\int d1\, d2\, \rho_1\rho_2\left(g_{12}\ln g_{12} - g_{12} + 1\right). \tag{72}$$

As we already know, $S_2 \leq 0$. From the inequality $\ln x \leq x - 1$, valid for all $x > 0$, we derive $-x\ln x \geq x - x^2$ for $x \geq 0$, and then obtain:
$$S_2 = \frac{1}{2}\int d1\, d2\, \rho_1\rho_2\left(-g_{12}\ln g_{12} + g_{12} - 1\right) \geq -\frac{1}{2}\int d1\, d2\, \rho_1\rho_2\left(g_{12} - 1\right)^2. \tag{73}$$

Clearly, estimating the size of the lower bound in Equation (73) is a much simpler problem than working with S_2 itself.

Taking $h_{12} \equiv g_{12} - 1$, it is evident that h_{12} shares all symmetries of g_{12}. Hence, we can write:
$$h_{12} = \sum_\mathbf{G} \widetilde{h}_\mathbf{G}(\mathbf{r}_1 - \mathbf{r}_2) e^{i\mathbf{G}\cdot\frac{\mathbf{r}_1+\mathbf{r}_2}{2}} \quad \text{with} \quad \widetilde{h}^*_\mathbf{G}(\mathbf{r}) = \widetilde{h}_{-\mathbf{G}}(\mathbf{r}) \text{ and } \widetilde{h}_\mathbf{G}(\mathbf{r}) = \widetilde{h}_\mathbf{G}(-\mathbf{r}). \tag{74}$$

Observe that the $\tilde{h}_\mathbf{G}(\mathbf{r})$ functions are nothing but Fourier coefficients, once the h function has been expressed in terms of $\mathbf{S} = (\mathbf{r}_1 + \mathbf{r}_2)/2$ and $\mathbf{r} = \mathbf{r}_1 - \mathbf{r}_2$:

$$\tilde{h}_\mathbf{G}(\mathbf{r}) = \frac{1}{v_0} \int_\mathcal{C} d^3 S\, h(\mathbf{S} + \mathbf{r}/2, \mathbf{S} - \mathbf{r}/2) e^{-i\mathbf{G}\cdot\mathbf{S}}. \tag{75}$$

By the Riemann–Lebesgue lemma, $\tilde{h}_\mathbf{G}(\mathbf{r}) \to 0$ as $G \to \infty$ (for arbitrary $\mathbf{r}$). Moreover, $\tilde{h}_\mathbf{G}(\mathbf{r}) \to 0$ for $r \to \infty$ (for arbitrary $\mathbf{G}$) since $\rho_{12} \to \rho_1 \rho_2$ for $|\mathbf{r}_1 - \mathbf{r}_2| \to \infty$. Similarly, for $k_{12} \equiv h_{12}^2$ we have that

$$k_{12} = \sum_\mathbf{G} \tilde{k}_\mathbf{G}(\mathbf{r}_1 - \mathbf{r}_2) e^{i\mathbf{G}\cdot\frac{\mathbf{r}_1+\mathbf{r}_2}{2}} \quad \text{with} \quad \tilde{k}_\mathbf{G}(\mathbf{r}) = \sum_{\mathbf{G}'} \tilde{h}_{\mathbf{G}-\mathbf{G}'}(\mathbf{r}) \tilde{h}_{\mathbf{G}'}(\mathbf{r}). \tag{76}$$

Now observe that, for $\rho^{(1)}(\mathbf{r}) = \sum_\mathbf{G} \tilde{u}_\mathbf{G} e^{i\mathbf{G}\cdot\mathbf{r}}$,

$$\rho^{(1)}(\mathbf{r}) \rho^{(1)}(\mathbf{r}') = \sum_\mathbf{G} \underbrace{\left(\sum_{\mathbf{G}'} \tilde{u}_{\mathbf{G}-\mathbf{G}'} \tilde{u}_{\mathbf{G}'} e^{i(2\mathbf{G}'-\mathbf{G})\cdot\frac{\mathbf{r}-\mathbf{r}'}{2}} \right)}_{\tilde{v}_\mathbf{G}^\infty(\mathbf{r}-\mathbf{r}')} e^{i\mathbf{G}\cdot\frac{\mathbf{r}+\mathbf{r}'}{2}}. \tag{77}$$

Using the above equation, and changing the integration variables from $\mathbf{r}_1$ and $\mathbf{r}_2$ to $\mathbf{S}$ and $\mathbf{r}$, we obtain:

$$-\frac{1}{2} \int d1\, d2\, \rho_1 \rho_2 (g_{12} - 1)^2 = -\frac{1}{2} \int d^3 S\, d^3 r \sum_\mathbf{G} \tilde{v}_\mathbf{G}^\infty(\mathbf{r}) e^{i\mathbf{G}\cdot\mathbf{S}} \sum_{\mathbf{G}'} \tilde{k}_{\mathbf{G}'}(\mathbf{r}) e^{i\mathbf{G}'\cdot\mathbf{S}}$$

$$= -\frac{1}{2} \sum_{\mathbf{G},\mathbf{G}'} \underbrace{\int d^3 S\, e^{i(\mathbf{G}+\mathbf{G}')\cdot\mathbf{S}}}_{V\delta_{\mathbf{G}',-\mathbf{G}}} \underbrace{\int d^3 r\, \tilde{v}_\mathbf{G}^\infty(\mathbf{r}) \tilde{k}_{\mathbf{G}'}(\mathbf{r})}_{i_{\mathbf{G},\mathbf{G}'}}$$

$$= -\frac{1}{2} V \sum_\mathbf{G} i_{\mathbf{G},-\mathbf{G}}, \tag{78}$$

where

$$i_{\mathbf{G},-\mathbf{G}} = \sum_{\mathbf{G}'} \tilde{u}_{\mathbf{G}-\mathbf{G}'} \tilde{u}_{\mathbf{G}'} \int d^3 r \left(\sum_{\mathbf{G}''} \tilde{h}_{-\mathbf{G}-\mathbf{G}''}(\mathbf{r}) \tilde{h}_{\mathbf{G}''}(\mathbf{r}) \right) e^{i(2\mathbf{G}'-\mathbf{G})\cdot\frac{\mathbf{r}}{2}}. \tag{79}$$

In the special case $\tilde{h}_\mathbf{G} = \mathcal{H}(\mathbf{r})\delta_{\mathbf{G},0}$, we have $h_{12} = \tilde{h}_0 = \mathcal{H}(\mathbf{r}_1 - \mathbf{r}_2)$ and $\tilde{k}_\mathbf{G}(\mathbf{r}) = \mathcal{H}(\mathbf{r})^2 \delta_{\mathbf{G},0}$. Then, from Equation (77) we derive

$$\sum_\mathbf{G} i_{\mathbf{G},-\mathbf{G}} = i_{0,0} = \int d^3 r\, \tilde{v}_0^\infty(\mathbf{r}) \tilde{k}_0(\mathbf{r}) = \sum_\mathbf{G} |\tilde{u}_\mathbf{G}|^2 \widetilde{\mathcal{H}^2}(\mathbf{G}). \tag{80}$$

An independent computation of the integral leads to the same result:

$$-\frac{1}{2} \int d1\, d2\, \rho_1 \rho_2 (g_{12} - 1)^2 = -\frac{1}{2} \int d^3 r\, \rho^{(1)}(\mathbf{r}) \int d^3 r'\, \underbrace{\rho^{(1)}(\mathbf{r}+\mathbf{r}')}_{\sum_\mathbf{G} \tilde{u}_\mathbf{G}^* e^{-i\mathbf{G}\cdot(\mathbf{r}+\mathbf{r}')}} \mathcal{H}^2(\mathbf{r}')$$

$$= -\frac{1}{2} \sum_\mathbf{G} \tilde{u}_\mathbf{G}^* \underbrace{\int d^3 r\, \rho^{(1)}(\mathbf{r}) e^{-i\mathbf{G}\cdot\mathbf{r}}}_{V\tilde{u}_\mathbf{G}} \underbrace{\int d^3 r'\, \mathcal{H}^2(\mathbf{r}') e^{-i\mathbf{G}\cdot\mathbf{r}'}}_{\widetilde{\mathcal{H}^2}(\mathbf{G})}$$

$$= -\frac{1}{2} V \sum_\mathbf{G} |\tilde{u}_\mathbf{G}|^2 \widetilde{\mathcal{H}^2}(\mathbf{G}), \tag{81}$$

which should be compared with Equation (66). For $\mathcal{H}(\mathbf{r}) = -1_\mathcal{C}(\mathbf{r})$ and $\tilde{u}_\mathbf{G} = \rho \exp\{-G^2/(4\alpha)\}$, we readily obtain $S_2 = -N/2$ from both Equations (66) and (78), meaning that in this case the two-body entropy coincides with its lower bound in Equation (73).

The quantity (80) is clearly $\mathcal{O}(1)$, since the summand is rapidly converging to zero; this implies that the two-body entropy of a crystal is, at least for $\tilde{h}_\mathbf{G} = \mathcal{H}(\mathbf{r})\delta_{\mathbf{G},0}$, bounded from below by a $\mathcal{O}(N)$ quantity. In the most general case, where Equations (78) and (79), rather, apply, we can only observe the following. As G grows in size, for any fixed $\mathbf{G}'$ and $\mathbf{G}''$ both $\tilde{u}_{\mathbf{G}-\mathbf{G}'}\tilde{u}_{\mathbf{G}'}$ and $\tilde{h}_{-\mathbf{G}-\mathbf{G}''}(\mathbf{r})\tilde{h}_{\mathbf{G}''}(\mathbf{r})$ get smaller, suggesting that $i_{\mathbf{G},-\mathbf{G}}$ will decrease too. However, this is not enough to conclude that $\sum_\mathbf{G} i_{\mathbf{G},-\mathbf{G}}$ is $\mathcal{O}(1)$, and the only way to settle the problem is numerical.

3.6. Numerical Evaluation of the Structure Functions

The utility of (36) clearly relies on the possibility of determining the integrand in simulation with sufficient accuracy. First we see how the one-body entropy, Equation (24), is computed. We start dividing V into a large number $M = V/v_c$ of identical cubes of volume v_c, chosen to be small enough that a cube contains the center of at most one particle. Let $c_\alpha = 0, 1$ (with $\alpha = 1, \ldots, M$) be the occupancy of the αth cube in a given system configuration and $\langle c_\alpha \rangle$ its canonical average as computed in a long Monte Carlo simulation of the weakly constrained crystal (to fix the center of mass of the crystal in space it is sufficient to keep one particle fixed; then, periodic boundary conditions will contribute to keep crystalline axes also fixed in the course of simulation). Given this setup, the local density at $\mathbf{r}_1$ (a point inside the αth cube) can be estimated as

$$\rho^{(1)}(\mathbf{r}_1) \approx \frac{\langle c_\alpha \rangle}{v_c}, \tag{82}$$

and the integral in (24) becomes

$$\int d^3 r_1\, \rho^{(1)}(\mathbf{r}_1) \ln \frac{\rho^{(1)}(\mathbf{r}_1)}{\rho} \approx \sum_{\alpha=1}^M \langle c_\alpha \rangle \ln \frac{\langle c_\alpha \rangle}{\rho v_c} \tag{83}$$

(notice that $\rho v_c = N/M \ll 1$; we need $v_c \to 0$ and an infinitely long simulation to make (82) an exact relation). Similarly, if $\mathbf{r}_2$ falls within the βth cube, then

$$\rho^{(2)}(\mathbf{r}_1, \mathbf{r}_2) \approx \frac{\langle c_\alpha c_\beta \rangle}{v_c^2} \tag{84}$$

and from Equation (30) we derive

$$\begin{aligned}
\tilde{g}(r) &\approx \frac{1}{\rho^2 V} \sum_{\alpha=1}^M v_c \frac{1}{N_\gamma} \sum_{|\gamma|=r} \frac{\langle c_\alpha c_{\alpha+\gamma} \rangle}{v_c^2} = \frac{1}{\rho^2 V} \left\langle \sum_{\alpha=1}^M v_c \frac{1}{N_\gamma} \sum_{|\gamma|=r} \frac{c_\alpha c_{\alpha+\gamma}}{v_c^2} \right\rangle \\
&= \frac{1}{M \rho^2 v_c^2} \left\langle \sum_{\alpha=1}^M \delta_{c_\alpha, 1} \frac{1}{N_\gamma} \sum_{|\gamma|=r} c_{\alpha+\gamma} \right\rangle.
\end{aligned} \tag{85}$$

In the above formula $N_\gamma \simeq 4\pi r^2 \Delta r / v_c$ is the number of cubes whose center lies at a distance r from α (to within a certain tolerance $\Delta r \ll r$), and the inner sum is carried out over those cubes only. Since $\rho v_c N_\gamma = 4\pi r^2 \Delta r \rho$ and $M \rho v_c = N$, an equivalent formula for $\tilde{g}(r)$ is

$$\tilde{g}(r) \approx \left\langle \frac{1}{N} \sum_{i=1}^N \frac{\mathcal{N}_i(r \pm \Delta r/2)}{4\pi r^2 \Delta r \rho} \right\rangle, \tag{86}$$

denoting $\mathcal{N}_i(r \pm \Delta r/2)$ the number of particles found at a distance between $r - \Delta r/2$ and $r + \Delta r/2$ from the ith particle in the given configuration. Equation (86) closely reflects the method of computing the radial distribution function in a CE simulation (see, e.g., equation (11) in reference [34]).

The function $\widetilde{g}(r)$ admits yet another expression, which further strengthens its resemblance to the $g(r)$ of a liquid (as reported e.g., in [35]). It follows from Equations (30) and (6) that

$$\rho^2 \widetilde{g}(r) = \frac{1}{V} \int \frac{d^2\Omega}{4\pi} \left\langle \sum_{ij}{}' \int d^3r_1 \, \delta^3(\mathbf{r}_1 - \mathbf{R}_i)\delta^3(\mathbf{r}_1 + \mathbf{r} - \mathbf{R}_j) \right\rangle. \tag{87}$$

Note that, for any sufficiently smooth function $f(\mathbf{r})$,

$$\begin{aligned}
\int d^3r \, f(\mathbf{r}) \int d^3r_1 \, \delta^3(\mathbf{r}_1 - \mathbf{R}_i)\delta^3(\mathbf{r}_1 + \mathbf{r} - \mathbf{R}_j) &= \int d^3r_1 \, \delta^3(\mathbf{r}_1 - \mathbf{R}_i) \int d^3r \, f(\mathbf{r}) \delta^3(\mathbf{r}_1 + \mathbf{r} - \mathbf{R}_j) \\
&= \int d^3r_1 \, \delta^3(\mathbf{r}_1 - \mathbf{R}_i) f(\mathbf{R}_j - \mathbf{r}_1) = f(\mathbf{R}_j - \mathbf{R}_i)
\end{aligned} \tag{88}$$

and

$$\begin{aligned}
\int d^3r \, f(\mathbf{r}) \int d^3r_1 \, \delta^3(\mathbf{r}_1 - \mathbf{R}_i)\delta^3(\mathbf{R}_i + \mathbf{r} - \mathbf{R}_j) &= \int d^3r_1 \, \delta^3(\mathbf{r}_1 - \mathbf{R}_i) \int d^3r \, f(\mathbf{r}) \delta^3(\mathbf{R}_i + \mathbf{r} - \mathbf{R}_j) \\
&= f(\mathbf{R}_j - \mathbf{R}_i) \underbrace{\int d^3r_1 \, \delta^3(\mathbf{r}_1 - \mathbf{R}_i)}_{1} = f(\mathbf{R}_j - \mathbf{R}_i),
\end{aligned} \tag{89}$$

we are allowed to replace $\delta^3(\mathbf{r}_1 - \mathbf{R}_i)\delta^3(\mathbf{r}_1 + \mathbf{r} - \mathbf{R}_j)$ with $\delta^3(\mathbf{r}_1 - \mathbf{R}_i)\delta^3(\mathbf{R}_i + \mathbf{r} - \mathbf{R}_j)$ in Equation (87), and thus obtain

$$\rho^2 \widetilde{g}(r) = \frac{1}{V} \int \frac{d^2\Omega}{4\pi} \left\langle \sum_{ij}{}' \delta^3(\mathbf{R}_i + \mathbf{r} - \mathbf{R}_j) \underbrace{\int d^3r_1 \, \delta^3(\mathbf{r}_1 - \mathbf{R}_i)}_{1} \right\rangle = \frac{1}{V} \int \frac{d^2\Omega}{4\pi} \left\langle \sum_{ij}{}' \delta^3(\mathbf{R}_i + \mathbf{r} - \mathbf{R}_j) \right\rangle, \tag{90}$$

which finally leads to

$$\rho \widetilde{g}(r) = \int \frac{d^2\Omega}{4\pi} \left\langle \frac{1}{N} \sum_i \sum_{j \neq i} \delta^3(\mathbf{R}_i + \mathbf{r} - \mathbf{R}_j) \right\rangle. \tag{91}$$

At zero temperature, we can neglect the average and simply write

$$\rho \widetilde{g}(r) = \int \frac{d^2\Omega}{4\pi} \frac{1}{N} \sum_i \sum_{j \neq i} \delta^3(\mathbf{R}_i + \mathbf{r} - \mathbf{R}_j) = \int \frac{d^2\Omega}{4\pi} \sum_{\mathbf{R} \neq 0} \delta^3(\mathbf{r} - \mathbf{R}) = \sum_{\mathbf{R} \neq 0} \frac{1}{4\pi R^2} \delta(r - R), \tag{92}$$

where in the last step we have followed the same path leading to Equation (41).

We can similarly proceed for the functions at Equations (31) and (35), which can be computed by the following formulae:

$$\widetilde{g}_0(r) \approx \frac{1}{\rho^2 V} \sum_{\alpha=1}^M v_c \frac{1}{N_\gamma} \sum_{|\gamma|=r} \frac{\langle c_\alpha \rangle \langle c_{\alpha+\gamma} \rangle}{v_c \, v_c} = \frac{1}{M\rho^2 v_c^2} \sum_{\alpha=1}^M \langle c_\alpha \rangle \frac{1}{N_\gamma} \sum_{|\gamma|=r} \langle c_{\alpha+\gamma} \rangle \tag{93}$$

and

$$\widetilde{h}(r) \approx \frac{1}{M\rho^2 v_c^2} \sum_{\alpha=1}^M \frac{1}{N_\gamma} \sum_{|\gamma|=r} \langle c_\alpha c_{\alpha+\gamma} \rangle \ln \frac{\langle c_\alpha c_{\alpha+\gamma} \rangle}{\langle c_\alpha \rangle \langle c_{\alpha+\gamma} \rangle}. \tag{94}$$

While $\widetilde{g}(r)$ is the statistical average of an estimator whose histogram can be updated in the course of the simulation (see Equation (86)), $\widetilde{g}_0(r)$ can only be estimated at the end of simulation, once $\langle c_\alpha \rangle$ has been evaluated for every α with effort comparable to that made for the one-body entropy. Much more costly is the calculation of $\widetilde{h}(r)$, which should also be performed at the end of simulation after evaluating $\langle c_\alpha c_\beta \rangle$ for every α and β.

Using translational lattice symmetry, the radial distribution functions and $\widetilde{h}(r)$ of a crystal can also be written as:

$$\rho \widetilde{g}(r) = \int_C d^3 r_1 \int \frac{d^2 \Omega}{4\pi} \rho^{(2)}(\mathbf{r}_1, \mathbf{r}_1 + \mathbf{r}); \quad \rho \widetilde{g}_0(r) = \int_C d^3 r_1 \int \frac{d^2 \Omega}{4\pi} \rho^{(1)}(\mathbf{r}_1) \rho^{(1)}(\mathbf{r}_1 + \mathbf{r});$$
$$\rho \widetilde{h}(r) = \int_C d^3 r_1 \int \frac{d^2 \Omega}{4\pi} \rho^{(2)}(\mathbf{r}_1, \mathbf{r}_1 + \mathbf{r}) \ln \frac{\rho^{(2)}(\mathbf{r}_1, \mathbf{r}_1 + \mathbf{r})}{\rho^{(1)}(\mathbf{r}_1) \rho^{(1)}(\mathbf{r}_1 + \mathbf{r})}, \tag{95}$$

leading to simplifying Equations (85), (93), and (94) into

$$\widetilde{g}(r) = \left\langle \sum_{\alpha=1}^{M/N} \delta_{c_\alpha,1} \frac{\sum_{|\gamma|=r} c_{\alpha+\gamma}}{4\pi r^2 \Delta r \rho} \right\rangle; \quad \widetilde{g}_0(r) = \sum_{\alpha=1}^{M/N} \langle c_\alpha \rangle \frac{\sum_{|\gamma|=r} \langle c_{\alpha+\gamma} \rangle}{4\pi r^2 \Delta r \rho};$$
$$\widetilde{h}(r) = \sum_{\alpha=1}^{M/N} \frac{1}{4\pi r^2 \Delta r \rho} \sum_{|\gamma|=r} \langle c_\alpha c_{\alpha+\gamma} \rangle \ln \frac{\langle c_\alpha c_{\alpha+\gamma} \rangle}{\langle c_\alpha \rangle \langle c_{\alpha+\gamma} \rangle}. \tag{96}$$

In the above formulae, the α index only runs over the cubes contained in a Wigner–Seitz/Voronoi cell of the lattice, while the β sum is still carried out over all cubes in the simulation box.

3.7. Numerical Tests

We first examine the shape of the structure functions $\widetilde{g}(r)$ and $\widetilde{g}_0(r)$ for hard spheres, choosing a r resolution of $\Delta r = 0.05$ (in units of the particle diameter σ). We take a system of $N = 4000$ particles arranged in a fcc lattice with packing fraction $\eta = 0.600$ (recall that the melting value is approximately 0.545). Periodic conditions are applied at the system boundary. In order to constrain the crystal in space, we keep one particle fixed during the simulation. As for $\widetilde{g}_0(r)$, we employ the Tarazona ansatz for $\alpha = 95$ (see Equation (39)), a value providing the best fit to the one-body density drawn from simulation.

We use the standard Metropolis Monte Carlo (MC) algorithm, constantly adjusting the maximum shift of a particle during equilibration until the fraction of accepted moves becomes close to 50% (then, the maximum shift is no longer changed). We produce 50,000 MC cycles in the equilibration run, whereas CE averages are computed over a total of further 2×10^5 cycles. Our results are plotted in Figure 1. While at short distances $\widetilde{g}(r)$ and $\widetilde{g}_0(r)$ are rather different, as r increases the oscillations of the two functions become closer and closer in amplitude.

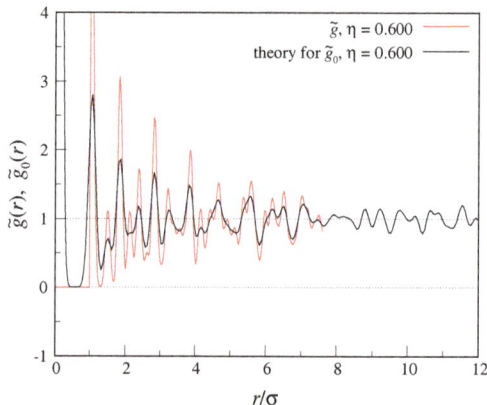

Figure 1. We show a comparison between $\widetilde{g}(r)$ for a fcc crystal of hard spheres ($\eta = 0.600$) and the $\widetilde{g}_0(r)$ function given in Equation (39), where the value of α (95) has been chosen such that the Tarazona ansatz (21) fits at best the one-body density drawn from simulation.

To obtain the one-body density with sufficient accuracy, we use a grid of about 50 points along each space direction in the unit cell. However, this grid resolution is too high for allowing the computation of $\tilde{h}(r)$, as the memory requirements for processing the $\langle c_\alpha c_\beta \rangle$ data are very huge. On the other hand, a coarser grid is incompatible with the Δr chosen.

To get closer to achieving our goal, i.e., to ascertain the N dependence of the two-body entropy for a crystal, we consider a two-dimensional system—hard disks. For this system, the transformation from fluid to solid occurs in two stages, via an intermediate hexatic fluid phase [36] (the transition from isotropic to hexatic fluid is first-order, whereas the hexatic-solid transition is continuous and occurs at $\eta = 0.700$). We consider a system of $N = 1152$ hard disks, arranged in a triangular crystal with packing fraction $\eta = 0.800$, and a mesh consisting of about 80 points along each direction in the unit cell. Even though translational correlations are only quasi-long-ranged in an infinite two-dimensional crystal, when one of the particles is kept artificially fixed this specificity is lost and the (finite) two-dimensional crystal is made fully similar to a three-dimensional crystal. Observe also that an infinite two-dimensional crystal shares at least the same breaking of rotational symmetry typical of an infinite three-dimensional crystal.

As before, we first look at the structure functions drawn from simulation, $\tilde{g}(r)$ and $\tilde{g}_0(r)$. Our results are plotted in Figure 2, together with the $\tilde{g}_0(r)$ function of Equation (40) for $\alpha = 75$. For this α the matching between the two $\tilde{g}_0$ functions is nearly perfect, indicating that the peaks of the one-body density are (to a high level of accuracy) Gaussian in shape. For $\eta = 0.800$ we find $S_1/N = -2.156$.

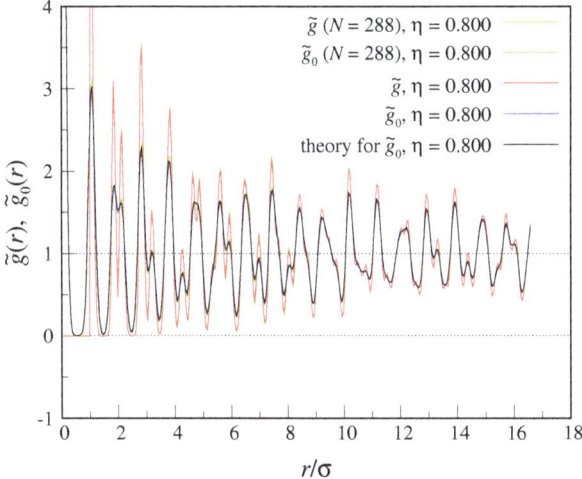

Figure 2. Structure functions $\tilde{g}(r)$ and $\tilde{g}_0(r)$ for a triangular crystal of hard disks ($\eta = 0.800$). We report data for two sizes, $N = 288$ and $N = 1152$. For comparison, we also plot the $\tilde{g}_0(r)$ function in Equation (40) for $\alpha = 75$. As is clear, the Tarazona ansatz represents an excellent model for the one-body density of the weakly-constrained hard-disk crystal.

In Figure 3, we show our main result, $\tilde{h}(r)$, for $\eta = 0.800$ and two different crystal sizes, $N = 288$ and 1152. We point out that, in order to obtain these data, we had to run a separate simulation for each r, as the memory usage is rather extreme. To be sure, we have computed the $\tilde{g}_0$ values in an independent way, i.e., using the same program loop written for $\tilde{h}(r)$, eventually finding the same results as in Figure 2. Looking at Figure 3, we see that $\tilde{h}(r)$ shows a series of peaks at neighbor positions and in the valleys within, taking preferentially positive values (meaning that its oscillations are not centered around zero). However, the damping of large-distance oscillations is too gradual to allow us to assess the nature of the asymptotic decay of $\tilde{h}(r)$ and then compute S_2. We must attempt a few explanations for this behavior of $\tilde{h}(r)$: On one hand, the decay of $\tilde{h}(r)$ may really be slow (at least in two dimensions), but S_2 would nonetheless be extensive, which implies a large S_2/N value. It may as well be that constraining the crystal in space by hinging the position of one particle has a strong effect on the speed of $\tilde{h}$ decay, which only a finite-size scaling of data can relieve. Indeed, when going from $N = 288$ to $N = 1152$ the values of $\tilde{h}$ are slightly shifted downwards.

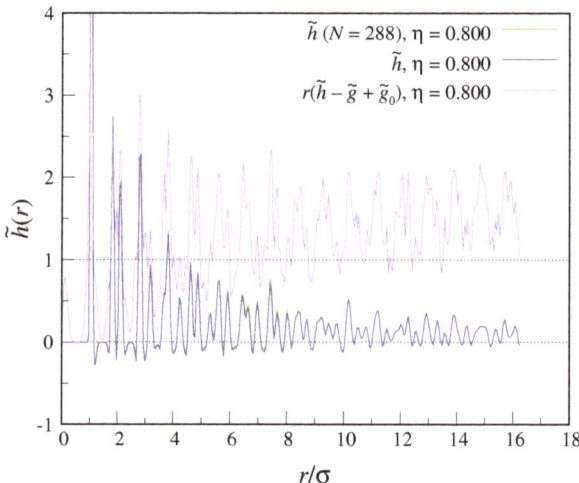

Figure 3. The function $\tilde{h}(r)$ for hard disks ($\eta = 0.800$). As in Figure 2, data for two sizes are shown, namely, $N = 288$ and $N = 1152$. It appears that the oscillations of $\tilde{h}(r)$ decay very slowly, which implies slow convergence of the integrand in Equation (36) to zero.

In summary, we have not reached any clear demonstration of S_2 extensivity in a crystal. This task has proved to be very hard to settle numerically. Our hope is that, based on our preparatory work, other authors with more powerful computational resources at their disposal can push the numerical analysis forward and eventually come up with a definite solution of the problem.

4. Conclusions

In this paper, we inquired into the possibility of extending the zero-RMPE criterion, a popular one-phase criterion of freezing for simple fluids, to also cover the melting of a solid. After revisiting the derivation of the entropy MPCE in the canonical ensemble, we argued that the formula applies for a crystal too. We exploited lattice symmetries to constrain the structures of one- and two-body densities, so as to gain as much information as possible on the first few terms in the entropy expansion. While this was enough to prove that the crystal one-body entropy is an extensive quantity, the information obtained was not sufficient to hold the same for the two-body entropy, whose scaling with the size of the crystal remains elusive. We thus attempted to clarify the question numerically, but we faced an insurmountable obstacle in the computational and memory limitations. To alleviate the problem, we turned towards a two-dimensional case, namely, hard disks, but with poor results: the structure function that must be integrated over distances to obtain the two-body entropy is weakly convergent to zero. In the near future, we plan to check whether the situation is more favorable for a different two-dimensional interaction, either endowed with an attractive tail (e.g., the Lennard-Jones potential) or provided with a soft core (for example, a Gaussian repulsion).

Author Contributions: Conceptualization, S.P. and P.V.G.; Methodology, S.P. and P.V.G.; Software, S.P.; Validation, S.P. and P.V.G.; Formal Analysis, S.P. and P.V.G.; Investigation, S.P. and P.V.G.; Resources, S.P. and P.V.G.; Data Curation, S.P.; Writing—Original Draft Preparation, S.P. and P.V.G.; Writing—Review & Editing, S.P. and P.V.G.; Visualization, S.P. and P.V.G.; Supervision, S.P. and P.V.G.; Project Administration, S.P. and P.V.G. All authors have read and agreed to the published version of the manuscript.

Funding: This research received no external funding.

Acknowledgments: This work has benefited from computer facilities made available by the PO-FESR 2007-2013 Project MedNETNA (Mediterranean Network for Emerging Nanomaterials).

Conflicts of Interest: The authors declare no conflict of interest.

Appendix A. Truncating the Entropy Expansion

We hereafter give an interpretation of the successive estimates of $S_N^{\text{exc}}/k_B = -\langle \ln P_{1...N} \rangle$ obtained by stopping the expansion (13) at a given order of correlations. We show that each truncated entropy expansion can be arranged in the form $-\langle \ln P_{1...N}^\star \rangle$, where $P_{1...N}^\star$ is a functional of all the MDFs up to n-th order, for $n = 1, 2, \ldots, N$ (however, without claiming that $P_{1...N}^\star$ represents a proper, i.e., normalized distribution). Our method resembles the one originally devised by H. S. Green to express the canonical entropy of a N-particle fluid in terms of correlation functions [1]. While H. S. Green correctly inferred the first three terms in the expansion, he did not provide a general recipe to obtain the further terms recursively.

For $N = 1$ there is only one MDF, $P_1 \equiv P^{(1)}(\mathbf{r}_1)$, in terms of which a fully symmetric approximation to $P_{1...N}$ can be constructed:

$$\text{1st} - \text{order approximation}: \quad P_{1...N}^\star = \prod_i^N P_i. \tag{A1}$$

Notice that $-\langle \ln P_{1...N}^\star \rangle = S_N^{(1)}/k_B$.

To obtain a better approximation we consider a system of two particles. Since

$$P_{12} = P_1 P_2 \times \frac{P_{12}}{P_1 P_2}, \tag{A2}$$

we see that P_{12} is the product of the 1st-order approximation (A1) times a correction factor $P_{12}/(P_1 P_2) = Q_{12}$. Assuming that in a N-particle system each distinct pair of particles contributes the same factor to $P_{1...N}^\star$, we arrive at the

$$\text{2nd} - \text{order approximation}: \quad P_{1...N}^\star = \prod_i^N P_i \prod_{i<j}^N Q_{ij}. \tag{A3}$$

The number of factors in the second product is $N(N-1)/2$. For this $P_{1...N}^\star$ we obtain

$$-\langle \ln P_{1...N}^\star \rangle = -N \int P_1 \ln P_1 - \binom{N}{2} \int P_{12} \ln Q_{12} = S_N^{(2)}/k_B. \tag{A4}$$

Moving to $N = 3$, we observe that

$$P_{123} = P_1 P_2 P_3 Q_{12} Q_{13} Q_{23} \times \frac{Q_{123}}{Q_{12} Q_{13} Q_{23}}, \tag{A5}$$

which is the second-order approximation to $P_{123}^\star$ times a correction factor. In the event that each distinct triplet of particles contributes the same factor to $P_{1...N}^\star$, we obtain the

$$\text{3rd} - \text{order approximation}: \quad P_{1...N}^\star = \prod_i^N P_i \prod_{i<j}^N Q_{ij} \prod_{i<j<k}^N \frac{Q_{ijk}}{Q_{ij} Q_{ik} Q_{jk}}. \tag{A6}$$

Notice that a different expression for the latter ratio is

$$\frac{Q_{ijk}}{Q_{ij}Q_{ik}Q_{jk}} = \frac{P_{ijk}}{\frac{P_{ij}P_{ik}P_{jk}}{P_i P_j P_k}}. \qquad (A7)$$

The number of factors in the third product is $N(N-1)(N-2)/6$. For this $P^\star_{1\ldots N}$, the approximate entropy is

$$-\langle \ln P^\star_{1\ldots N}\rangle = -N\int P_1 \ln P_1 - \binom{N}{2}\int P_{12}\ln Q_{12} - \binom{N}{3}\left[\int P_{123}\ln Q_{123} - \binom{3}{2}\int P_{12}\ln Q_{12}\right] = \frac{S^{(3)}_N}{k_B}. \qquad (A8)$$

We can similarly proceed to derive higher-order approximations. The 4-body MDF of a system of $N=4$ particles is trivially decomposed as

$$P_{1234} = P_1 P_2 P_3 P_4 Q_{12} Q_{13} Q_{14} Q_{23} Q_{24} Q_{34} \frac{Q_{123}}{Q_{12}Q_{13}Q_{23}} \frac{Q_{124}}{Q_{12}Q_{14}Q_{24}} \frac{Q_{134}}{Q_{13}Q_{14}Q_{34}} \frac{Q_{234}}{Q_{23}Q_{24}Q_{34}}$$
$$\times \frac{Q_{1234}}{Q_{12}Q_{13}Q_{14}Q_{23}Q_{24}Q_{34}\frac{Q_{123}}{Q_{12}Q_{13}Q_{23}}\frac{Q_{124}}{Q_{12}Q_{14}Q_{24}}\frac{Q_{134}}{Q_{13}Q_{14}Q_{34}}\frac{Q_{234}}{Q_{23}Q_{24}Q_{34}}}$$
$$= P_1 P_2 P_3 P_4 Q_{12}Q_{13}Q_{14}Q_{23}Q_{24}Q_{34} \frac{Q_{123}}{Q_{12}Q_{13}Q_{23}}\frac{Q_{124}}{Q_{12}Q_{14}Q_{24}}\frac{Q_{134}}{Q_{13}Q_{14}Q_{34}}\frac{Q_{234}}{Q_{23}Q_{24}Q_{34}}$$
$$\times \frac{Q_{1234}}{\frac{Q_{123}Q_{124}Q_{134}Q_{234}}{Q_{12}Q_{13}Q_{14}Q_{23}Q_{24}Q_{34}}}, \qquad (A9)$$

whence the

$$4\text{th} - \text{order approximation}: \quad P^\star_{1\ldots N} = \prod_i^N P_i \prod_{i<j}^N Q_{ij} \prod_{i<j<k}^N \frac{Q_{ijk}}{Q_{ij}Q_{ik}Q_{jk}} \prod_{i<j<k<l}^N \frac{Q_{ijkl}}{\frac{Q_{ijk}Q_{ijl}Q_{ikl}Q_{jkl}}{Q_{ij}Q_{ik}Q_{il}Q_{jk}Q_{jl}Q_{kl}}}. \qquad (A10)$$

Notice that a different expression for the latter ratio is

$$\frac{Q_{ijkl}}{\frac{Q_{ijk}Q_{ijl}Q_{ikl}Q_{jkl}}{Q_{ij}Q_{ik}Q_{il}Q_{jk}Q_{jl}Q_{kl}}} = \frac{P_{ijkl}}{\frac{P_{ijk}P_{ijl}P_{ikl}P_{jkl}}{\frac{P_{ij}P_{ik}P_{il}P_{jk}P_{jl}P_{kl}}{P_i P_j P_k P_l}}}. \qquad (A11)$$

For this $P^\star_{1\ldots N}$ we obtain

$$-\langle \ln P^\star_{1\ldots N}\rangle = -N\int P_1 \ln P_1 - \binom{N}{2}\int P_{12}\ln Q_{12} - \binom{N}{3}\left[\int P_{123}\ln Q_{123} - \binom{3}{2}\int P_{12}\ln Q_{12}\right]$$
$$- \binom{N}{4}\left[\int P_{1234}\ln Q_{1234} - \binom{4}{3}\int P_{123}\ln Q_{123} + \binom{4}{2}\int P_{12}\ln Q_{12}\right] = S^{(4)}_N/k_B. \qquad (A12)$$

Eventually, with the last Nth-order approximation we recover the exact distribution, namely, $P^\star_{1\ldots N} = P_{1\ldots N}$, and the full entropy. Notice that, except for $n=1$ and N, the nth-order functional $P^\star_{1\ldots N}$ is not normalized.

We now provide a formalization of the procedure sketched above. For each value of n and each grouping $I_n = \{i_1, i_2, \ldots i_n\}$ of n particle indices, we write $P(I_n) \equiv P_{i_1\ldots i_n}$ as a product of positive cumulant factors to be determined recursively, that is

$$P(I_n) = \prod_{S_1 \subset I_n} C(S_1) \cdots \prod_{S_{n-1} \subset I_n} C(S_{n-1}) \times C(I_n), \qquad (A13)$$

where $\prod_{S_k \subset I_n}$ indicates the product over all k-tuples of distinct entries from I_n—there are $\binom{n}{k}$ factors in the product $\prod_{S_k \subset I_n}$. As shown before, $C(\{i\}) = P_i$, $C(\{i,j\}) = P_{ij}/(P_i P_j)$, $C(\{i,j,k\}) = P_{ijk} P_i P_j P_k/(P_{ij} P_{ik} P_{jk})$, and so on. Taking the logarithm of (A13) we obtain:

$$\ln P(I_n) = \ln C(I_n) + \sum_{S_{n-1} \subset I_n} \ln C(S_{n-1}) + \ldots + \sum_{S_1 \subset I_n} \ln C(S_1), \tag{A14}$$

which can be solved with respect to cumulants by the Möbius inversion formula (see, e.g., Equations (3) and (4) of reference [5]):

$$\ln C(I_n) = \ln P(I_n) - \sum_{S_{n-1} \subset I_n} \ln P(S_{n-1}) + \ldots + (-1)^{n-1} \sum_{S_1 \subset I_n} \ln P(S_1), \tag{A15}$$

which is in turn equivalent to writing

$$C(I_n) = \cfrac{P(I_n)}{\cfrac{\prod_{S_{n-1} \subset I_n} P(S_{n-1})}{\vdots \atop \prod_{S_1 \subset I_n} P(S_1)}}. \tag{A16}$$

Equations (A7) and (A11) are just particular cases of the above formula, respectively for $n = 3$ and $n = 4$. Once the cumulants have been determined, the functional $P^*_{1 \ldots N}$ of M-th order (for $M = 1, \ldots, N$) can be written, by an obvious change of notation, as

$$P^*_{1 \ldots N} = \prod_i^N C_i \prod_{i<j}^N C_{ij} \cdots \prod_{i_1 < \ldots < i_M}^N C_{i_1 \ldots i_M}, \tag{A17}$$

leading to

$$-\langle \ln P^*_{1 \ldots N} \rangle = -N \int P_1 \ln C_1 - \binom{N}{2} \int P_{12} \ln C_{12} - \ldots - \binom{N}{M} \int P_{12 \ldots M} \ln C_{12 \ldots M}. \tag{A18}$$

In view of Equation (A16), the above quantity is nothing but $S_N^{(M)}$.

References

1. Green, H.S. *The Molecular Theory of Fluids*; North Holland: Amsterdam, The Netherlands, 1952; pp. 70–73.
2. Nettleton, R.E.; Green, M.S. Expression in Terms of Molecular Distribution Functions for the Entropy Density in an Infinite System. *J. Chem. Phys.* **1958**, *29*, 1365–1370. [CrossRef]
3. Baranyai, A.; Evans, D.J. Direct entropy calculation from computer simulation of liquids. *Phys. Rev. A* **1989**, *40*, 3817–3822. [CrossRef] [PubMed]
4. Schlijper, A.G. Convergence of the cluster-variation method in the thermodynamic limit. *Phys. Rev. B* **1983**, *27*, 6841–6848. [CrossRef]
5. An, G. A Note on the Cluster Variation Method. *J. Stat. Phys.* **1988**, *52*, 727–734. [CrossRef]
6. Pelizzola, A. Cluster variation method in statistical physics and probabilistic graphical models. *J. Phys. A* **2005**, *38*, R309–R339. [CrossRef]
7. Hernando, J.A. Thermodynamic potentials and distribution functions: I. A general expression for the entropy. *Mol. Phys.* **1990**, *69*, 319–326. [CrossRef]
8. Prestipino, S.; Giaquinta, P.V. Statistical entropy of a lattice-gas model: Multiparticle correlation expansion. *J. Stat. Phys.* **1999**, *96*, 135–167; Erratum: *ibid* **2000**, *98*, 507–509. [CrossRef]
9. Prestipino, S.; Giaquinta, P.V. The entropy multiparticle-correlation expansion for a mixture of spherical and elongated particles. *J. Stat. Mech. Theor. Exp.* **2004**, P09008. [CrossRef]
10. D'Alessandro, M. Multiparticle correlation expansion of relative entropy in lattice systems. *J. Stat. Mech. Theor. Exp.* **2016**, *2016*, 073201. [CrossRef]

11. Maffioli, L.; Clisby, N.; Frascoli, F.; Todd, B.D. Computation of the equilibrium three-particle entropy for dense atomic fluids by molecular dynamics simulation. *J. Chem. Phys.* **2019**, *151*, 164102. [CrossRef]
12. Abramo, M.C.; Caccamo, C.; Costa, D.; Giaquinta, P.V.; Malescio, G.; Munaò, G.; Prestipino, S. On the determination of phase boundaries via thermodynamic integration across coexistence regions. *J. Chem. Phys.* **2015**, *142*, 214502. [CrossRef] [PubMed]
13. Giaquinta, P.V.; Giunta, G. About entropy and correlations in a fluid of hard spheres. *Phys. A* **1992**, *187*, 145–158. [CrossRef]
14. Giaquinta, P.V.; Giunta, G.; Prestipino Giarritta, S. Entropy and the freezing of simple liquids. *Phys. Rev. A* **1992**, *45*, R6966–R6968. [CrossRef] [PubMed]
15. Saija, F.; Pastore, G.; Giaquinta, P.V. Entropy and Fluid-Fluid Separation in Nonadditive Hard-Sphere Mixtures. *J. Phys. Chem. B* **1998**, *102*, 10368–10371. [CrossRef]
16. Donato, M.G.; Prestipino, S.; Giaquinta, P.V. Entropy and multi-particle correlations in two-dimensional lattice gases. *Eur. Phys. J. B* **1999**, *11*, 621–627. [CrossRef]
17. Saija, F.; Prestipino, S.; Giaquinta, P.V. Entropy, correlations, and ordering in two dimensions. *J. Chem. Phys.* **2000**, *113*, 2806–2813. [CrossRef]
18. Costa, D.; Micali, F.; Saija, F.; Giaquinta, P.V. Entropy and Correlations in a Fluid of Hard Spherocylinders: The Onset of Nematic and Smectic Order. *J. Phys. Chem. B* **2002**, *106*, 12297–12306. [CrossRef]
19. Prestipino, S. Analog of surface preroughening in a two-dimensional lattice Coulomb gas. *Phys. Rev. E* **2002**, *66*, 021602. [CrossRef]
20. Saija, F.; Saitta, A.M.; Giaquinta, P.V. Statistical entropy and density maximum anomaly in liquid water. *J. Chem. Phys.* **2003**, *119*, 3587–3589. [CrossRef]
21. Speranza, C.; Prestipino, S.; Malescio, G.; Giaquinta, P.V. Phase behavior of a fluid with a double Gaussian potential displaying waterlike features. *Phys. Rev. E* **2014**, *90*, 012305. [CrossRef]
22. Prestipino, S.; Malescio, G. Characterization of the structural collapse undergone by an unstable system of ultrasoft particles. *Phys. A* **2016**, *457*, 492–505. [CrossRef]
23. Banerjee, A.; Nandi, M.K.; Sastry, S.; Bhattacharyya, S.M. Determination of onset temperature from the entropy for fragile to strong liquids. *J. Chem. Phys.* **2017**, *147*, 024504. [CrossRef] [PubMed]
24. Santos, A.; Saija, F.; Giaquinta, P.V. Residual Multiparticle Entropy for a Fractal Fluid of Hard Spheres. *Entropy* **2018**, *20*, 544. [CrossRef]
25. Frenkel, D. Order through entropy. *Nat. Mater.* **2015**, *14*, 9–12. [CrossRef]
26. Speedy, R.J. The entropy of a glass. *Mol. Phys.* **1993**, *80*, 1105–1120. [CrossRef]
27. Berthier, L.; Ozawa, M.; Scalliet, C. Configurational entropy of glass-forming liquids. *J. Chem. Phys.* **2019**, *150*, 160902. [CrossRef]
28. Baus, M.; Tejero, C.F. *Equilibrium Statistical Physics*; Springer: Berlin, Germany, 2008; pp. 61–63.
29. Tarazona, P. A density functional theory of melting. *Mol. Phys.* **1984**, *52*, 81–96. [CrossRef]
30. Prestipino, S.; Giaquinta, P.V. Ground state of weakly repulsive soft-core bosons on a sphere. *Phys. Rev. A* **2019**, *99*, 063619. [CrossRef]
31. Prestipino, S.; Sergi, A.; Bruno, E. Freezing of soft-core bosons at zero temperature: A variational theory. *Phys. Rev. B* **2018**, *98*, 104104. [CrossRef]
32. Rascón, C.; Mederos, L.; Navascués, G. Thermodynamic consistency of the hard-sphere solid distribution function. *J. Chem. Phys.* **1996**, *105*, 10527–10534. [CrossRef]
33. Gernoth, K.A. Spatial Microstructure of Quantum Crystals. *J. Low. Temp. Phys.* **2002**, *126*, 725–730. [CrossRef]
34. Prestipino Giarritta, S.; Ferrario, M.; Giaquinta, P.V. Statistical geometry of hard particles on a sphere: Analysis of defects at high density. *Phys. A* **1993**, *201*, 649–665. [CrossRef]
35. Hansen, J.-P.; McDonald, I.R. *Theory of Simple Liquids*; Academic: Oxford, UK, 2013.
36. Bernard, E.P.; Krauth, W. Two-Step Melting in Two Dimensions: First-Order Liquid-Hexatic Transition. *Phys. Rev. Lett.* **2011**, *107*, 155704. [CrossRef] [PubMed]

© 2020 by the authors. Licensee MDPI, Basel, Switzerland. This article is an open access article distributed under the terms and conditions of the Creative Commons Attribution (CC BY) license (http://creativecommons.org/licenses/by/4.0/).

Article

Structural and Thermodynamic Peculiarities of Core-Shell Particles at Fluid Interfaces from Triangular Lattice Models

Vera Grishina [1], Vyacheslav Vikhrenko [1] and Alina Ciach [2,*]

[1] Department of Mechanics and Engineering, Belarusian State Technological University, 13a Sverdlova Str., 220006 Minsk, Belarus; vera1grishina@gmail.com (V.G.); vvikhre@gmail.com (V.V.)
[2] Institute of Physical Chemistry, Polish Academy of Sciences, Kasprzaka 44/52, 01-224 Warszawa, Poland
* Correspondence: aciach@ichf.edu.pl

Received: 12 September 2020; Accepted: 22 October 2020; Published: 26 October 2020

Abstract: A triangular lattice model for pattern formation by core-shell particles at fluid interfaces is introduced and studied for the particle to core diameter ratio equal to 3. Repulsion for overlapping shells and attraction at larger distances due to capillary forces are assumed. Ground states and thermodynamic properties are determined analytically and by Monte Carlo simulations for soft outer- and stiffer inner shells, with different decay rates of the interparticle repulsion. We find that thermodynamic properties are qualitatively the same for slow and for fast decay of the repulsive potential, but the ordered phases are stable for temperature ranges, depending strongly on the shape of the repulsive potential. More importantly, there are two types of patterns formed for fixed chemical potential—one for a slow and another one for a fast decay of the repulsion at small distances. In the first case, two different patterns—for example clusters or stripes—occur with the same probability for some range of the chemical potential. For a fixed concentration, an interface is formed between two ordered phases with the closest concentration, and the surface tension takes the same value for all stable interfaces. In the case of degeneracy, a stable interface cannot be formed for one out of four combinations of the coexisting phases, because of a larger surface tension. Our results show that by tuning the architecture of a thick polymeric shell, many different patterns can be obtained for a sufficiently low temperature.

Keywords: core-shell particles; liquid interfaces; triangular lattice; thermodynamics; ground states; structure; line tension; phase coexistence; competing interaction; fluctuations

1. Introduction

Metal or semiconducting nanoparticles find numerous applications in catalysis, optics, biomedicine, environmental science, and so forth. In order to prevent the charge-neutral nanoparticles from aggregation, recently, various types of core-shell nanoparticles (CSNPs) have been produced [1,2]. In the CSNPs, the hard, typically metal or semiconducting nanoparticle with a diameter ranging from a few tens to a few hundreds of nanometers is covered by a soft polymeric shell. The polymeric chains can interpenetrate, and the distance between the particles can become smaller than the shell diameter at some energetic cost. This energetic cost, or the softness of the shells, can be controlled in particular by the crosslinking of the polymeric chains. The shell-to-core diameter ratio in the majority of the experiments varies from about 1.1 to about 4 [2–5]. Since the shell thickness can be controlled independently of the core diameter, and the effective interactions between the CSNPs depend on the thickness and the architecture of the shells, the desired effective interactions can be obtained by choosing different protocols for the synthesis.

Properties of the CSNPs have been intensively studied not only in the bulk, but also at the fluid interfaces [1–12]. It turned out that while in the bulk, the effective interaction consists of the hard core followed by the soft repulsive shoulder; at fluid interfaces, strong capillary attraction can be present for separations larger than the shell diameter, in addition to the above mentioned repulsive interactions at shorter distances [2–4].

Experimental results show that the CSNPs at fluid interfaces can form interesting patterns, depending on the properties of the particles and on the fraction of the interface area covered by them [2–4]. For a small area fraction, highly ordered arrays of hexagonally packed particles are typically observed. Compression may lead to the sudden formation of particle clusters [1]. The surface pressure–area isotherms can have a characteristic shape of alternating segments with a very large and quite small slope, and the large compressibility signals structural changes [2,3]. The origin of these patterns and of the structural changes, their nature, and dependence on the properties of the CSNPs are not fully understood yet. The theory of CSNPs, that at fluid interfaces, they repel each other at short separations and attract each other at large separations, is much less developed than the experimental studies [13–15]. This is in contrast to theoretical and simulation studies of patterns formed by particles with soft repulsive potentials [16].

Because there are many factors controlling the core-and-shell diameter and the architecture of the shell, there is a need for a simplified, coarse-grained theory that could predict general trends in pattern formation for various ranges, strengths, and shapes of the effective potential. In Ref. [13], a one-dimensional (1D) lattice model with repulsion between nearest neighbors and attraction between second or third neighbors was solved exactly. The obtained isotherms consist of alternating segments with very large and quite small slopes, as in experiments. The steep parts of the isotherms are associated with periodic patterns. The number of the steps, however, is larger in the case of third-neighbor attraction (i.e., thicker shell), and depends on the strengths of the repulsive and attractive parts of the potential. There are no phase transitions in the thermodynamic sense in 1D, but the correlation function shows oscillatory decay with the correlation length that for a strong attraction can be very large (10^4 times the core diameter). These results show a strong dependence of the structure and mechanical properties of the monolayers of the CSNPs on the range, shape, and strength of the effective interactions and agree with the experimental observation of the more complex behavior of the CSNPs with thick shells. However, any 1D model cannot answer the question if different patterns correspond to different phases, and obviously only 1D patterns can be examined.

Particles with a size equal to or larger than a few tens of nanometers are practically irreversibly adsorbed at the interface, but can move freely in the interface area [2]. For this reason, the particles trapped at the interface can be modeled as a two-dimensional system. Since closely packed CSNPs form a hexagonal pattern, triangular lattice models with the lattice constant a equal to the diameter of the hard core (or the distance of the closest approach of the particles upon compression) are appropriate and convenient generic models for CSNPs at fluid interfaces. In Ref. [15], lattice models for CSNPs with thin and thick shells were introduced and studied. Following Ref. [16], we assumed that the shell-to-core ratio separating the thin and thick shells is $\sqrt{3}$. According to this criterion, the shell is thin when the shells of the second neighbors of closely-packed particle-cores do not overlap, otherwise it is thick.

For thin shells, nearest-neighbor repulsion and second neighbor attraction between particle cores occupying sites of the triangular lattice were assumed. The shell-to-core ratio in this model is $\sqrt{3}$. We have found four phases in this model—very dilute gas, hexagonal lattice of closely packed shells, hexagonal lattice of vacancies, and closely packed cores. We have calculated the surface tension between coexisting phases for different orientations of the interface and found that the particles at the stable interfaces corresponding to the smallest surface tension lie on straight lines. The interface lines meet at the angles 60° or 120°. When the fixed area fraction of the CSNPs is smaller than the area fraction of the hexagonal phase, large voids with a hexagonal shape are formed. The results are in good agreement with the experiment in Reference [2–4].

In order to study the effect of the shell thickness, in the second model, we assume that the inner shell is covered by a much softer outer shell, and the nearest-neighbor repulsion is followed by vanishing interactions for the second, third, and fourth neighbors and by attraction between the fifth neighbors. The shell-to-core ratio in this model is equal to 3, as in the experiments of Refs. [1,3]. Six more phases, including honeycomb lattices of particles or vacancies and periodically ordered rough clusters were found, but these additional phases were stable only at the coexistence with the phases found for the thin-shell model. For the fixed area fraction of the particles, two phases with the closest area fraction to the mean one, and the interface between them, were present at low temperature. For increasing T, islands of different phases in the sea of the hexagonally packed shells were observed in the course of simulations for the area fraction exceeding the value at the close packing of the shells. Such complex patterns, somewhat similar to the patterns observed in experiments [2–4], occur because of the metastability of several ordered phases and large interface fluctuations in 2D.

The results of Reference [15] concern CSNPs with composite shells with a stiff inner shell and very soft outer shell. In this work, we focus on the question of the role of the shape of the repulsive shoulder, associated with the architecture of the crosslinked polymeric chains. We assume first- and second-neighbor repulsion, and fifth-neighbor attraction, and consider different second- to first-neighbor repulsion ratios. The model is introduced in Section 2. The ground state of an open system and of the system with a fixed number of particles is determined in Sections 3.1 and 3.2, respectively. We find the same patterns as in Ref. [15] for weak second-neighbor repulsion, but the patterns absent for thin shells are present for some intervals of the chemical potential. For the second-to-first neighbor repulsion ratio larger than 1/3, the stable patterns are completely different. Moreover, for some ranges of the chemical potential, the ground state is degenerated, and two quite different patterns are stable. In Section 4, thermodynamic properties obtained for $T > 0$ by Monte Carlo simulations are described. We present the chemical potential, compressibility, and specific heat as functions of the concentration. Section 5 contains our conclusions.

2. The Model

The system that models the core-shall particles on a surface is described in Ref. [15]. The thermodynamic Hamiltonian of the system is:

$$H^* = \frac{1}{2} \sum_{k=1}^{k_{max}} \sum_{k_i=1}^{z_k} \sum_{i=1}^{M} J_k^* \hat{n}_i \hat{n}_{k_i} - \mu^* \sum_{i=1}^{M} \hat{n}_i, \qquad (1)$$

where k_i numerates the sites of the k-th coordination sphere around the site i, z_k is the coordination number, J_k^* is the interaction constant for the k-th coordination sphere, $\hat{n}_i$ is the occupation number (0 or 1), and μ^* is the chemical potential. The particles can occupy sites of a triangular lattice containing $M = L \times L$ lattice sites.

The lattice parameter a is equal to the diameter of the hard core of the particles. It is supposed that the particles repel each other with the intensity $J_1^* = J_1 J$ ($J_1 = 1$), if the particles occupy the nearest neighbor sites, and feel weaker repulsion on the next nearest sites with the intensity $J_2^* = J_2 J$. The intermediate third and fourth neighbors do not interact ($J_3 = J_4 = 0$), while the fifth neighbors attract each other with the energy $J_5^* = -J_5 J$. Thus, J_2 and J_5 are the dimensionless interaction energies ($J_2 = J_2^*/J$, $J_5 = -J_5^*/J$, and J has units of energy). The dimensionless chemical potential $\mu = \mu^*/J$ and dimensionless temperature $T = k_B T^*/J$ will be used as well. In the terminology of Reference [15], it is model II with an additional repulsive interaction of the second neighbors.

The interaction potential as a function of the distance between the particle cores is shown in Figure 1. As is demonstrated below, the variation of the second and fifth neighbor interactions can lead to different symmetry-breaking (heterostructural) transitions in the system; it was recently attained using the augmented potential [17]. Compared to conventional interaction potentials that are determined a priori, the augmented potentials adjust the effective interactions on the basis of the local environment of each particle and efficiently capture multi-body effects at a local level.

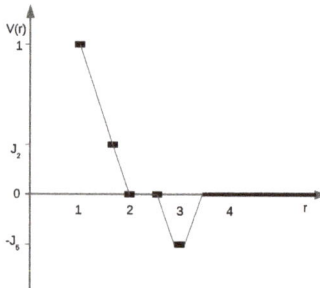

Figure 1. The interaction potential as a function of the distance between the particle centers in units of the first-neighbor repulsion J. The symbols denote the interaction between the cores occupying the lattice sites, and the line is to guide the eye. The shown interactions $J_2 = J_5 = 1/3$ correspond to a crossover between different patterns formed by the particles, as described in Section 3.1.

In accordance with the range of the interparticle interactions (up to the fifth neighbors), the unit cell contains nine (3 × 3) lattice sites. In this case, the fifth neighbors (that correspond to the largest interaction range) belong to the nearest unit cells and the distance between them determines the translation vector that preserves the symmetry of the system. The subsequent calculation (Section 3) and simulation (Section 4) shows that such a choice accounts for all possible ordered states of the system. For describing the ordered states, the lattice is decomposed in nine sublattices (Figure 2).

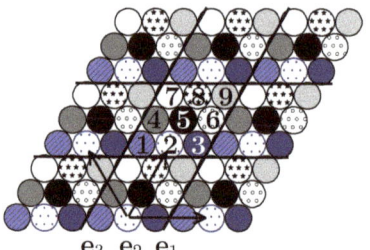

Figure 2. The system of unit cells with particles belonging to nine sublattices and the lattice vectors $e_i, i = 1, 2, 3$. The particles 1 and 2 or 1 and 4 are the nearest neighbors, the particles 1 and 5 or 2 and 7 are the next nearest neighbors, the particles 1 and 3 or 1 and 7 are the third neighbors, the particles 1 and 6 are the fourth neighbors. The particles with the same texture in nearest unit cells with the separation $3a$ are the fifth neighbors.

3. The Ground States

3.1. The Ground States of Open Systems

In the system with repulsive interactions of the first and second neighbors and attraction of the fifth neighbors, the ground states with ten concentrations $n/9$, $n = 0, 1, ..., 9$ with a different distribution of the particles over the unit cell are possible. At zero temperature, the stable configurations are determined by the minima of the dimensionless thermodynamic Hamiltonian per lattice site

$$\omega = H^*/MJ \qquad (2)$$

because the entropy does not contribute to the thermodynamic functions.

In the vacuum state $\omega_0 = 0$. In the $c = 1/9$ phase, each particle has six neighbors of the fifth order. Calculating the system energy, each interacting bond is taken into account twice. Thus,

$$\omega_{1/9} = (-3J_5 - \mu)/9. \qquad (3)$$

For $c = 2/9$, two possibilities exist for the distribution of two particles over the unit cell (Figure 3). The calculated potentials are as follows:

$$\begin{aligned}\text{(a)} \quad & \omega_{2/9} = (3J_2 - 6J_5 - 2\mu)/9, \\ \text{(c)} \quad & \omega_{2/9} = (1 - 6J_5 - 2\mu)/9.\end{aligned} \qquad (4)$$

the 1 in Equation (4)c, originates from the nearest neighbor interaction for this configuration.

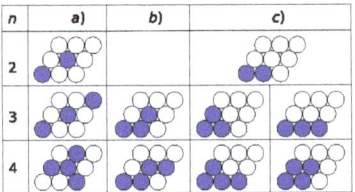

Figure 3. The possible distributions of particles over the unit cell for the concentrations $c = n/9$ with $2 \leq n \leq 4$. For the concentrations with $5 \leq n \leq 7$, the particles and vacancies have to be interchanged. For the concentrations with $n = 1$ or 8, the particle or vacancy can occupy any lattice site of the unit cell. There are several equivalent distributions of particles over the unit cell for the concentrations with $2 \leq n \leq 4$ or vacancies for the concentrations with $5 \leq n \leq 7$.

For larger concentrations, we can write the expressions corresponding to the columns a–c in Figure 3

$$\begin{aligned}\text{(a)} \quad & \omega_{3/9} = (9J_2 - 9J_5 - 3\mu)/9, \\ \text{(b)} \quad & \omega_{3/9} = (2 + 3J_2 - 9J_5 - 3\mu)/9, \\ \text{(c)} \quad & \omega_{3/9} = (3 - 9J_5 - 3\mu)/9,\end{aligned} \qquad (5)$$

$$\begin{aligned}\text{(a)} \quad & \omega_{4/9} = (3 + 9J_2 - 12J_5 - 4\mu)/9, \\ \text{(b)} \quad & \omega_{4/9} = (4 + 6J_2 - 12J_5 - 4\mu)/9, \\ \text{(c)} \quad & \omega_{4/9} = (5 + 3J_2 - 12J_5 - 4\mu)/9.\end{aligned} \qquad (6)$$

For the concentration $c = n/9$ with $n \geq 5$, the distribution of the vacancies in the unit cell is the same as the distribution of the particles for $c = (9-n)/9$. In the dense state ($n = 9$, $c = 1$), all the lattice sites are filled by the particles. The ω for these states can be calculated as

$$\omega_{n/9} = \omega_{1-n/9} + (2n-9)[3(1+J_2-J_5) - \mu]/9, \quad 5 \leq n \leq 9 \qquad (7)$$

for each particular distribution of the particles/vacancies over the unit cell.

Comparing the r.h.s. of Equations (4)–(6), we can see that at $J_2 < 1/3$ the system states shown in the column (a) of Figure 3 are more stable, while at $J_2 > 1/3$, the system states shown in the column (c) are preferable. The system states of the column (b) could occur if additional interactions of the third and/or fourth neighbors were taken into account.

Thus, the presence of the interaction between the second neighbors eliminates the phase degeneration observed in the system without it [15] and results in additional stable phases in certain regions of the chemical potential.

To make further analysis more transparent, we consider the case $J_5 = J_2 \equiv J_{2,5}$. There exists the crossover value of the interaction parameter $J_{2,5} = 1/3$, which separates the possible states of the system. Let us consider particular values of the interaction parameter, below and above the crossover, $J_{2,5} = 1/4 < 1/3$ and $J_{2,5} = 1/2 > 1/3$, that correspond to the slower and faster decay of the interaction potential for short separations as compared with $V(r)$, shown in Figure 1, respectively.

In Figure 4, the grand thermodynamic potentials per lattice site, Equations (3)–(7), are shown as functions of the chemical potential at $J_{2,5} = 1/4$. The stable states correspond to the lowest value of ω for given μ, i.e., to the lowest line segments between the intersection points. These segments determine the chemical potential intervals corresponding to particular concentrations of the particles. Each intersection point corresponds to the coexistence of two phases with the closest concentrations. The stable system states and the corresponding chemical potential intervals are shown in Figure 5. The phase diagram of the system at $J_{2,5} < 1/3$ is shown in Figure 6.

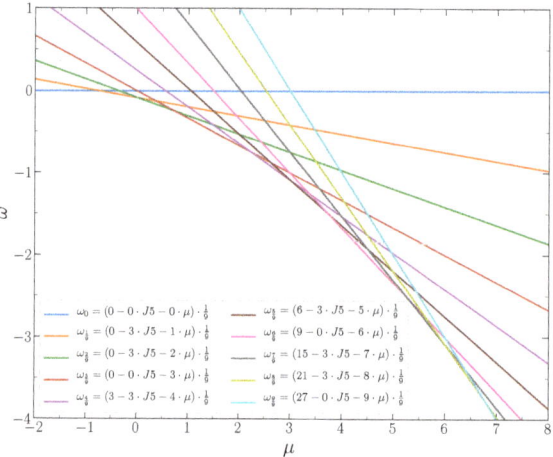

Figure 4. The dimensionless thermodynamic Hamiltonian per lattice site versus the chemical potential for the concentrations $c = n/9, n = 0, 1, 2, ...9$ at $J_{2,5} = 1/4$.

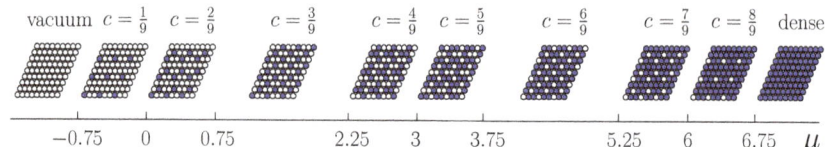

Figure 5. Cartoons of the distribution of particles over the lattice sites and the corresponding chemical potential intervals for the system with $J_{2,5} = 1/4$.

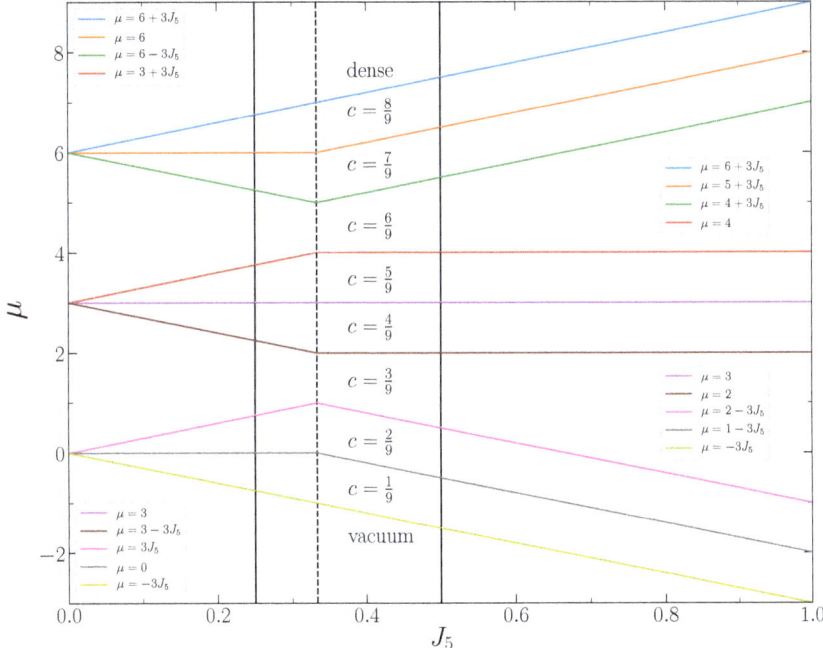

Figure 6. The phase diagram of the system. The vertical lines show the chemical potential intervals for the system states at $J_{2,5} = 1/4$, $J_{2,5} = 1/3$ and $J_{2,5} = 1/2$.

The phase coexistence lines between the phases with the concentrations $(n-1)/9$ and $n/9$ can be represented by the expression

$$\mu_{k,l} = 3k + 3(l-1)J_{2,5}, \quad k = 0, 1, 2; \quad l = 0, 1, 2,$$
$$n = 1 + 3k + l. \tag{8}$$

The phase diagram of the system for the larger interaction parameter, $J_{2,5} > 1/3$, is shown in Figure 6 as well. In this case, the phase coexistence lines between the phases with concentrations $(n-1)/9$ and $n/9$ obey the expression

$$\mu_{k,l} = k + 2l + 3(l-1)J_{2,5}, \quad k = 0, 1, 2; \quad l = 0, 1, 2,$$
$$n = 1 + k + 3l. \tag{9}$$

The dependence of the potential ω on the chemical potential at $J_{2,5} > 1/3$ is shown in Figure 7. The stable states correspond to the lowest line segments between the intersection points, which indicate the chemical potential values for the phase coexistence.

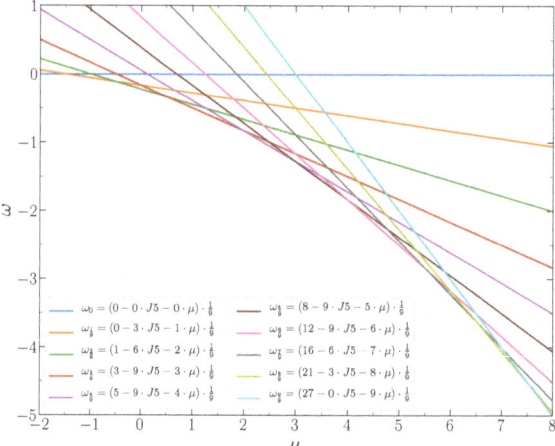

Figure 7. The dimensionless thermodynamic Hamiltonian per lattice site versus the chemical potential for the concentrations $c = n/9, n = 0, 1, 2, ..., 9$ at $J_{2,5} = 1/2$.

The structure of the stable phases at $J_{2,5} > 1/3$ is shown in Figure 8. At the concentrations $3/9$, $4/9, 5/9$, and $6/9$, the degenerated ground states exist. E.g., either triangles of the nearest neighbors or stripes parallel to the lattice vectors can exist with equal probabilities at $c = 3/9$. Ordered rhombuses or stripes with additional particles attached to them can occur at the concentration $c = 4/9$.

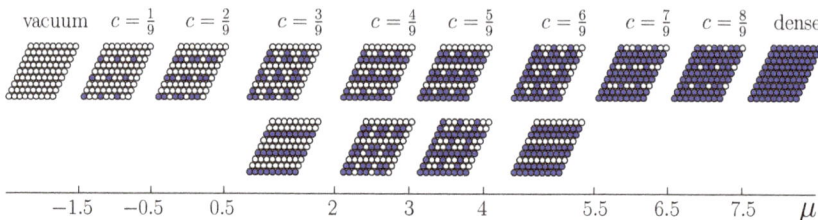

Figure 8. Cartoons of the distribution of particles over the lattice sites and the corresponding chemical potential intervals for the system with $J_{2,5} > 1/3$.

At the crossover value of the parameter $J_{2,5} = 1/3$, all the system states for the concentration between $2/9$ and $7/9$ are degenerated. The $\omega_{n/9}$ functions of the chemical potential μ according to Equations (3)–(7) coincide for all possible system states for a given value of n. Any distribution of particles shown in columns a) and c) of Figure 3 can occur at this value of the parameter $J_{2,5}$ on the line segments between the intersection points at $\mu = n - 2$ for the coexisting phases $c = (n-1)/9$ and $c = n/9, n = 1, 2, ..., 9$.

3.2. The Ground States for Fixed Number of Particles

In systems with a fixed number of particles, an arbitrary mean concentration can be considered. At $c \neq n/9$, two phases separated by an interface line coexist. Like in the previous case [15], the interface lines can be parallel or perpendicular to the lattice vectors $\mathbf{e}_i, i = 1, 2, 3$ (Figure 3). We have verified that the interface lines parallel to the lattice vectors are preferable because their line tensions are smaller.

On the coexistence line between vacuum and the $c = 1/9$ phase, each particle of the latter phase loses two interacting bonds with the fifth neighbors in the vacuum phase. The distance between particles along the interface is equal $3a$, where a is the nearest neighbor distance. Thus, the line tension is equal to $\sigma = J_{2,5}/3a$. At the interface line perpendicular to the lattice vectors $\mathbf{e_i}$, each particle in the first row loses three interacting bonds and in the second row one interacting bond. The distance between particles along the interface line is equal to $3a\sqrt{3}$. The line tension is $2J_{2,5}/3a\sqrt{3} > J_{2,5}/3a$. Thus, the interface lines between the coexisting phases $c = 0$ and $c = 1/9$ are parallel to the lattice vectors. The same conclusion, $\sigma = J_{2,5}/3a$, follows for the interface line between the phases $c = 1/9$ and $c = 2/9$ in both cases $J_{2,5} < 1/3$ and $J_{2,5} > 1/3$. Figure 9 demonstrates that one of the particles in the near interface unit cell of the $c = 2/9$ phase has six neighbors of the fifth order, while the other one has four such neighbors. The minimum of the line tension is assumed when one of the particles of the $c = 2/9$ phase in the unit cell has two fifth neighbors and no first and second neighbors in the phase $c = 1/9$. We finally note that the interface is twofold degenerated for the case of $J_{2,5} > 1/3$ at the concentrations $3/9 \leq c \leq 6/9$ (see Figure 8).

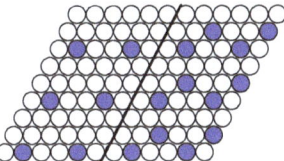

Figure 9. Cartoon of the interface between the $c = 1/9$ and $2/9$ phases for the system at $J_{2,5} < 1/3$.

As an example of a structure in a system with a fixed number of particles, the simulation snapshot for the system of 37 particles on the lattice of 36×36 lattice sites (the concentration $c = 0.029$) is shown in Figure 10. In the ideal case of $T = 0$, the particles have to form a regular hexagon. However, the simulation was done at a quite low but nonzero temperature $T = 0.1$. As a result, the coexistence of the rarefied gas phase (the vacuum state with a few evaporated particles as defects) and the phase with $c = 1/9$ is obtained. The interface lines are parallel to the lattice vectors in agreement with the analytical calculation for $T = 0$.

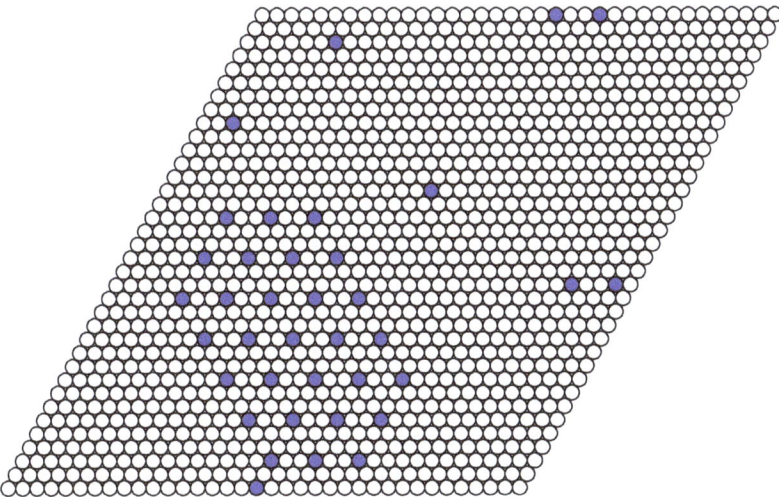

Figure 10. The snapshot of the system of 37 particles on the lattice of 36×36 lattice sites ($c \simeq 0.029$) after 9 000 Monte Carlo simulation steps (MCS) at $J_{2,5} = 1/2$. The interface lines are parallel to the lattice vectors $\mathbf{e_i}$.

At $J_{2,5} < 1/3$, the system states for a subsequent concentration can be produced from the previous one by adding a particle in the unit cell. At the interface line, this particle has no counterparts in the lower concentration phase, losing two interacting bonds with the fifth neighbors and the line tension is equal $\sigma = J_{2,5}/3a$ for all the coexisting phases with concentrations $(n-1)/9$ and $n/9$ for $n = 1, 2, ..., 9$.

The situation at $J_{2,5} > 1/3$ is more complicated. The neighboring system states in the upper row in Figure 3 differ from each other by one particle in the unit cell and the line tension for all the interfaces parallel to the lattice vectors is equal to $J_{2,5}/3a$. The system state for the $c = 3/9$ phase in the lower row can coexist with the $c = 2/9$ phase, because this state differs from the previous phase by one particle as well. Thus, the degeneracy of the phase coexistence between the phases $c = 2/9$ and $c = 3/9$ is observed.

However, the $c = 4/9$ phase in the lower row in Figure 3 can coexist with the $c = 3/9$ phase of the upper row, but not with that of the lower row, because they differ by positions of more than one particle. Thus, three combinations of coexisting phases exist at the mean concentration $3/9 < c < 4/9$. The same situation exists for the coexistence of the $c = 4/9$ and $c = 5/9$ phases. The states of the upper row can coexist between themselves as well as with the cross phases of the lower row. However, these phases in the lower row cannot form a stable interface between themselves. There are no counterparts for two particles in the more concentrated phase as well as for one particle of the less concentrated phase. The line tension in this case is three times larger. Three combinations of coexisting phases separated by a stable interface exist at the mean concentration $4/9 < c < 5/9$ as well.

The system states are symmetric with respect to $\mu = 3$ or $c = 0.5$ and the particle-vacancy interchange [18]. The phase coexistence at larger chemical potentials and concentrations are symmetric to their lower values.

4. Thermodynamics of the System at $T > 0$

At low dimensionless temperatures $T = k_B T^*/J$ (where T^* is the absolute temperature and k_B the Boltzmann constant), the ordered states found in the ground state remain present in the system, while the ordering is destroyed gradually with the temperature increase, due to thermal fluctuations. In this section, the Monte Carlo (MC) simulation results for $\mu(c)$ isotherms, isothermal compressibility, and specific heat are presented for the system at two values of the interaction parameter $J_{2,5} = 1/4$ and $J_{2,5} = 1/2$ below and above the crossover value $J_{2,5} = 1/3$. The Metropolis importance sampling simulations were performed with the chemical potential step $\Delta\mu = 0.02$ for the system of 96×96 lattice sites with periodic boundary conditions. one thousand Monte Carlo simulation steps (MCS) were used for equilibration. The subsequent 10 000 MCS were used for calculating the average values.

The isotherms displaying the concentration dependence on the chemical potential at $J_{2,5} = 1/2$ demonstrate typical behavior at low temperatures $T = 0.1, 0.2,$ and 0.3 (Figure 11). The wide empty horizontal segments in the $\mu(c)$ plots indicate forbidden regions of concentrations. These two-phase segments are separated by very steep parts of the $\mu(c)$ plots, where μ increases rapidly for very narrow range of c, centered at $n/9$. These intervals of c separating the horizontal segments expand with increasing temperature. The concentration intervals around $n/9, n = 0, 1, ..., 9$ correspond to the ordered patterns discussed in the previous section in the case of the grand canonical ensemble. At larger temperature $T = 0.5$, the horizontal regions in the $\mu(c)$ plot almost disappear. Thus, the critical temperature can be estimated as $T_{cr} \simeq 0.6$. The concentration increases continuously as a function of the chemical potential at this temperature. The repulsion interaction between the second neighbors not only removes the degeneracy of the system states at the concentrations $1/3$ and $2/3$, but also significantly reduces the critical temperature, which is around 1.1 when the second neighbors do not interact [15].

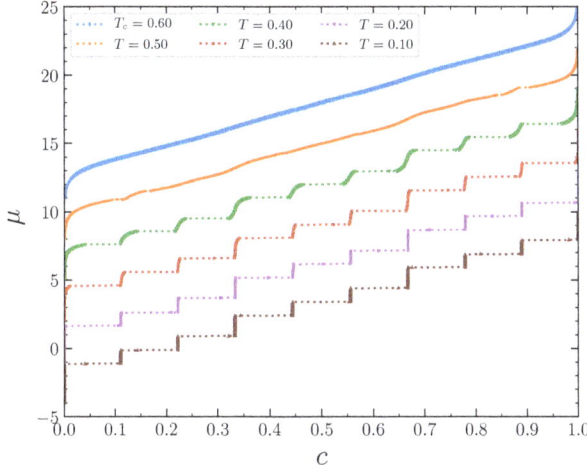

Figure 11. The chemical potential as a function of the concentration at $J_{2,5} = 0.5$ and several temperatures. The isotherms are shifted in the vertical direction by 3 from each other for clarity. The isotherm at $T = 0.1$ is not shifted.

The structure of stable phases resembles the ground state configurations. In Figures 12–14 the snapshots of the system at the concentration close to 4/9 are shown as examples. At the interaction parameter $J_{2,5} > 1/2$ above its critical value $1/3$, the system is degenerated and in different runs the final state Figures 13 and 14 corresponds to the possible ground state configurations given in Figure 8 with a few defects, due to thermal fluctuations.

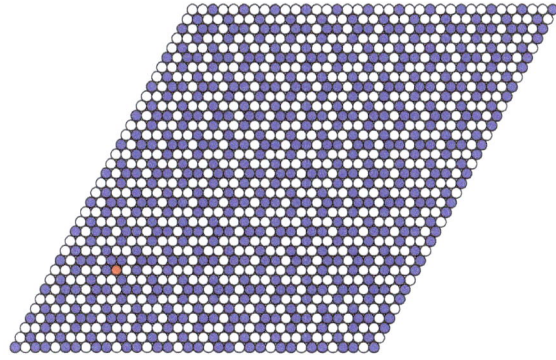

Figure 12. Snapshot of the system at $T = 0.2$, $\mu = 2.6$, $J_{2,5} = 1/4$ after 8 000 MCS. The extra particle (defect) is shown in red. This structure corresponds to the ground state configuration shown in Figure 5.

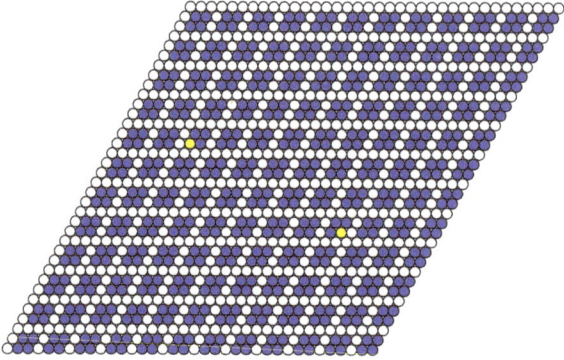

Figure 13. Snapshot of the system at $T = 0.3$, $\mu = 2.6$, $J_{2,5} = 1/2$ after 8 000 MCS. The additional vacancies (defects) are shown in yellow. This structure corresponds to the ground state configuration shown in the bottom row of Figure 8.

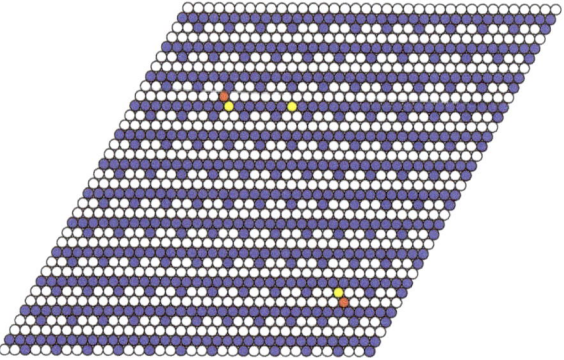

Figure 14. Snapshot of the system at $T = 0.3$, $\mu = 2.6$, $J_{2,5} = 1/2$ after 8 000 MCS. The extra particles and additional vacancies (defects) are shown in red and yellow, respectively. This structure corresponds to the ground state configuration shown in the upper row of Figure 8.

The phase transitions can be more clearly revealed by considering fluctuations (Figure 15). The inverse value of the thermodynamic factor $\chi_T = c(\partial(\beta\mu)/\partial c)_T$ is proportional to the concentration fluctuations

$$\chi_T^{-1} = \frac{\langle (N - \langle N \rangle)^2 \rangle}{\langle N \rangle} \tag{10}$$

that in turn is proportional to the isothermal compressibility $\kappa_T = (\partial c/\partial p)_T/c$, $\chi_T^{-1} = cT\kappa_T$, where the angular brackets $\langle ... \rangle$ denote averaging over the grand canonical ensemble, N is the number of particles in the system, $c = \langle N \rangle / M$ is the mean lattice concentration, and p is the pressure.

At the lowest temperatures, $T = 0.1$ and $T = 0.2$, the concentration fluctuations of each phase exist in narrow concentration regions. At the temperature $T = 0.4$, the minima of the concentration fluctuations are well distinguishable. They are attained at the most ordered system states with concentrations equal to a multiple 1/9. At $T = 0.5$, the concentration fluctuations span almost over the entire concentration range (0,1), indicating the approach to the critical temperature.

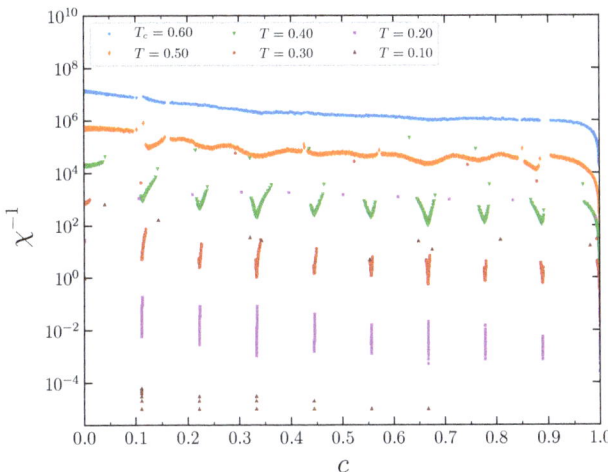

Figure 15. The inverse thermodynamic factor as a function of the concentration at $J_{2,5} = 0.5$ and several temperatures. The curves are shifted in the vertical direction by 3^{3n} from the lowest one for clarity.

Similar behavior is observed for the dimensionless specific heat (Figure 16), which is proportional to the energy fluctuations

$$c_V = \frac{1}{k_B \langle N \rangle} \left(\frac{\partial E^*}{\partial T^*} \right)_\mu = \frac{\langle (E - \langle E \rangle)^2 \rangle}{\langle N \rangle T^2}, \quad (11)$$

where $E = E^*/J$ is the dimensionless system energy (see the first term on the RHS in Equation (1)), and

$$\langle E \rangle = \frac{1}{2} \left\langle \sum_{k=1}^{5} \sum_{k_i=1}^{z_k} \sum_{i=1}^{M} J_k \hat{n}_i \hat{n}_{k_i} \right\rangle. \quad (12)$$

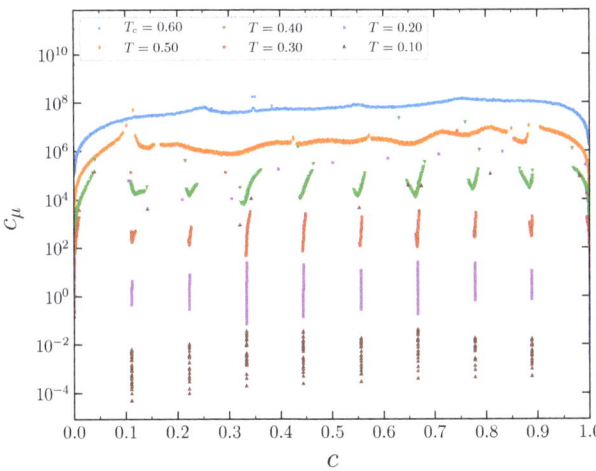

Figure 16. The dimensionless specific heat as a function of the concentration at $J_{2,5} = 0.5$ and several temperatures. The curves are shifted in the vertical direction by 3^{3n} from the lowest one for clarity.

As an example, the fine structure of the concentration and energy fluctuations is shown in Figure 17 and simulated with the reduced chemical potential step. The minima of these characteristics are close to the concentration $c = 1/3$ of the most ordered system state.

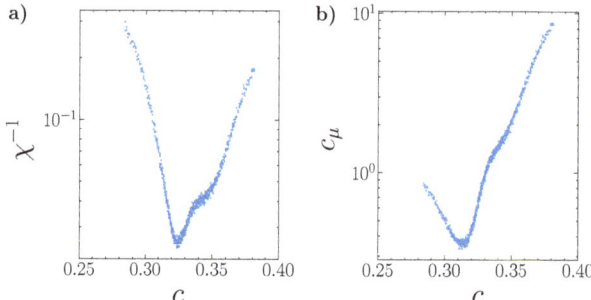

Figure 17. The fine structure of the inverse thermodynamic factor (**a**) and dimensionless specific heat (**b**) at $T = 0.3$ and $J_{2,5} = 0.5$ simulated with the reduced chemical potential step $\Delta \mu = 0.002$. The scatter of the results characterizes the precision of the simulation.

Structural peculiarities of the system can be tracked by considering the order parameters. The occupancy of particular sublattices (Figure 2) represents nine such order parameters. At the lowest temperatures, $T = 0.1$ and 0.2, the succession of the order parameters is in fact represented by the step functions rising from 0 to 1 when the concentration attains the value equal to a multiple of $1/9$. The order parameters become smoother when the temperature increases.

For the interaction parameter below its critical value $1/3$, in particular for $J_{2,5} = 1/4$, the chemical potential isotherms look like in the previous case (Figure 18), but the critical temperature is even lower, around 0.4. At low temperatures, the inverse thermodynamic factor and the specific heat, as well as the order parameters, have the same prominent features at the concentrations around a multiple of $1/9$. The rapid changes of the above quantities for $c \approx n/9$ are smoothed out with the temperature increase.

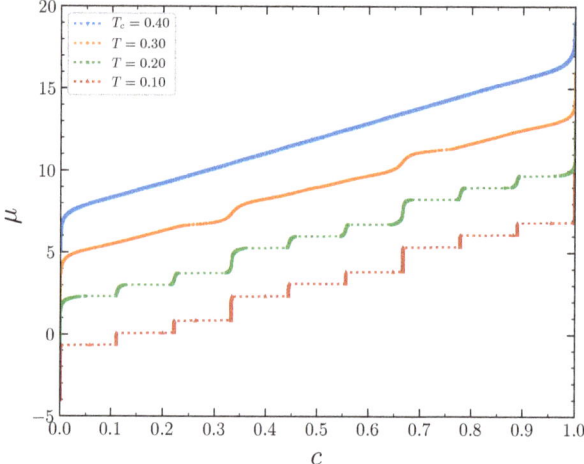

Figure 18. The chemical potential as a function of the concentration at $J_{2,5} = 0.25$ and several temperatures. The isotherms are shifted in the vertical direction by 3 from each other for clarity. The isotherm at $T = 0.1$ is not shifted.

Despite very similar thermodynamic properties of the systems with the interaction parameter $J_{2,5}$ above and below its crossover value $J_{2,5} = 1/3$, the structural characteristics are completely different above and below the crossover. The particle distribution over the lattice sites corresponds to the structures of Figure 5 or Figure 8 in the former and in the latter case, respectively. Due to thermal fluctuations at non-zero temperatures, the structures contain defects, which are particles on the sites that do not belonging to the ideal configurations or vacant sites on these configurations. The number of defects increases with the temperature increase, and the concentration range corresponding to the ordered phases increases as well.

5. Conclusions

We considered pattern formation by particles with hard cores covered by thick polymeric shells on a fluid interface. The structure of the shell can be designed in experimental studies, and it determines the effective repulsion between the particles. For this reason, the aim of our study was the determination of the effect of the shape of the repulsive potential on the patterns formed by the particles. We focused on the question how the rate at which the repulsion decreases with the distance influences the pattern formation. We considered a triangular lattice model with the lattice sites occupied by the particle cores. The nearest and next nearest neighbors repel each other, due to the overlapping polymeric shells, while the fifth neighbors attract each other because of the capillary forces.

The second neighbor repulsion in addition to the nearest and fifth neighbor interaction results in the significant enrichment of the ground state structures. Alongside with the concentrations equal to a multiple of $1/3$, the states with the concentrations equal to a multiple of $1/9$ are present for certain intervals of the chemical potential.

Importantly, the patterns formed for the concentration $c = n/9, n = 2,...,7$, are completely different for the repulsion that shows a fast and a slow decrease with the interparticle separation (see Figures 5 and 8 for the first and the second case, respectively). In the second case, two quite different patterns can occur with the same probability for given μ. The crossover value of the second-neighbor repulsion (equal to the fifth neighbor attraction) separating the two types of patterns is $J_{2,5} = 1/3$. Our model is suitable for thick composite shells, with a stiff inner part and soft outer part. The results show that by modifying the thickness and the structure of the stiff inner part, we can obtain completely different patterns on an interface.

In systems with a fixed number of particles, the energetically preferable interface lines are parallel to the lattice vectors for all stable interfaces. We have shown that in the case of the degenerated ground states, two types of patterns with lower density can coexist with two patterns with higher density, giving together four pairs of coexisting patterns. A stable interface, however, cannot be formed for one of these pairs.

At non-zero but not too large temperatures, the system passes through the structures corresponding to the ground states with thermally initiated defects. At low temperatures, the stable states exist in very narrow concentration intervals close to the concentrations equal to a multiple of $1/9$. The concentration intervals enlarge with the temperature increase. At temperatures slightly above the critical one, the concentration versus chemical potential isotherms become continuous. The critical temperature depends strongly on the shape of the interaction potential too. In the case of $J_{2,5} = 1/2$, i.e., when the intensity of the second neighbor repulsion is equal to the fifth neighbor attraction and twice as low as the first neighbor repulsion, the critical temperature is almost two times lower than in the system with vanishing interaction between the second neighbors [15]. The critical temperature decreases with decreasing interaction between the second and the fifth neighbors.

The fluctuations of the number of particles and energy are maximal at the concentrations corresponding to the phase transition points, and minimal in the most ordered states at concentrations close to a multiple of $1/9$. The order parameters determined as the mean concentrations on the sublattices demonstrate fast increase from 0 to 1, while the mean system concentration crosses a value equal to a multiple of $1/9$.

Quantitative comparison of the predictions of our model with experimental results is not possible yet. Due to the sensitive dependence of the patterns formed by the particles on the interfaces on the details of the effective interactions, it is necessary to precisely design and control the crosslinking architecture of the polymeric shell, in order to obtain the desired shape of the effective interactions in experiment. On the theoretical side, it would be necessary to compute the pressure–area fraction isotherms that are measured in experiments. On the qualitative level, however, our predictions for the chemical potential—area fraction isotherms can be compared with the experimental pressure–area fraction isotherms for the CSNPs with the shell-to-core size ratio ~ 3, since the chemical potential and the pressure play similar roles. Indeed, steps in the isotherms are clearly seen in Ref. [3], in qualitative agreement with our results. Moreover, structural evolution for increasing area fraction, in particular the formation of clusters, and the orientation of the interface between coexisting phases in our theory and in experiment are similar.

To summarize, we stress that our model indicates that by careful construction of the polymeric shell, one should be able to obtain core-shell nanoparticles forming a variety of different patterns on an interface.

Author Contributions: Conceptualization and methodology, A.C. and V.V.; software and simulation, V.G.; writing–original draft preparation, V.G. and V.V.; writing–review and editing, A.C.; data analysis, V.G., V.V., and A.C. All authors have read and agreed to the published version of the manuscript.

Funding: This project has received funding from the European Union Horizon 2020 research and innovation under the Marie Skłodowska-Curie grant agreement No 734276 (CONIN). An additional support in the years 2017–2020 has been granted for the CONIN project by the Polish Ministry of Science and Higher Education.

Conflicts of Interest: The authors declare no conflict of interest.

References

1. Vasudevan, S.A.; Rauh, A.; Barbera, L.; Karg, M.; Isa, L. Stable in bulk and aggregating at the interface: Comparing core–shell nanoparticles in suspension and at fluid interfaces. *Langmuir* **2018**, *34*, 886. [CrossRef] [PubMed]
2. Isa, L.; Buttinoni, I.; Fernandez-Rodriguez, M.A.; Vasudevan, S.A. Two-dimensional assemblies of soft repulsive colloids confined at fluid interfaces. *Europhys. Lett.* **2017**, *119*, 26001. [CrossRef]
3. Rauh, A.; Rey, M.; Barbera, L.; Zanini, M.; Karg, M.; Isa, L. Compression of hard core–soft shell nanoparticles at liquid–liquid interfaces: influence of the shell thickness. *Soft Matter* **2017**, *13*, 158. [CrossRef] [PubMed]
4. Rey, M.; Elnathan, R.; Ditcovski, R.; Geisel, K.; Zanini, M.; Fernandez-Rodriguez, M.A.; Naik, V.V.; Frutiger, A.; Richtering, W.; Ellenbogen, T.; et al. Fully tunable silicon nanowire arrays fabricated by soft nanoparticle templating. *Nano Lett.* **2016**, *16*, 157. [CrossRef] [PubMed]
5. Riest, J.; Mohanty, P.; Schurtenberger, P.; Likos, C.N. Coarse-graining of ionic microgels: Theory and experiment. *Z. Phys. Chem.* **2012**, *226*, 711. [CrossRef]
6. Vogel, N.; Fernandez-Lopez, C.; Perez-Juste, J.; Liz-Marzan, L.M.; Landfester, K.; Weiss, C.K. Ordered arrays of gold nanostructures from interfacially assembled Au@ PNIPAM hybrid nanoparticles. *Langmuir* **2012**, *28*, 8985. [CrossRef] [PubMed]
7. Nazli, K.O.; Pester, C.; Konradi, A.; Boker, A.; van Rijn, P. Cross-Linking Density and Temperature Effects on the Self-Assembly of SiO_2—PNIPAAm Core–Shell Particles at Interfaces. *Chemistry* **2013**, *19*, 5586. [CrossRef] [PubMed]
8. Volk, K.; Fitzgerald, J.P.; Retsch, M.; Karg, M. Time-Controlled Colloidal Superstructures: Long-Range Plasmon Resonance Coupling in Particle Monolayers. *Adv. Mater.* **2015**, *27*, 7332. [CrossRef] [PubMed]
9. Honold, T.; Volk, K.; Rauh, A.; Fitzgerald, J.P.S.; Karg, M. Tunable plasmonic surfaces via colloid assembly. *J. Mater. Chem. C* **2015**, *3*, 11449. [CrossRef]
10. Geisel, K.; Rudov, A.A.; Potemkin, I.I.; Richtering, W. Hollow and core–shell microgels at oil–water interfaces: Spreading of soft particles reduces the compressibility of the monolayer. *Langmuir* **2015**, *31*, 13145. [CrossRef] [PubMed]
11. Karg, M. Functional materials design through hydrogel encapsulation of inorganic nanoparticles: Recent developments and challenges. *Macromol. Chem. Phys.* **2016**, *217*, 242. [CrossRef]

12. Sheverdin, A.; Valagiannopoulos, C. Core-shell nanospheres under visible light: Optimal absorption, scattering, and cloaking. *Phys. Rev. B* **2019**, *99*, 075305. [CrossRef]
13. Ciach, A.; Pekalski, J. Exactly solvable model for self-assembly of hard core–soft shell particles at interfaces. *Soft Matter* **2017**, *13*, 2603. [CrossRef] [PubMed]
14. Groda, Ya.G.; Grishina, V.; Ciach, A.; Vikhrenko V.S. Phase diagram of the lattice fluid with SRLA-potential on the plane triangular lattice. *J. Belarusian State Univ. Phys.* **2019**, *3*, 81. [CrossRef]
15. Grishina, V.; Vikhrenko, V.; Ciach, A. Triangular lattice models for pattern formation by core-shell particles with different shell thicknesses. *J. Phys. Condens. Matter* **2020**, *32*, 405102. [CrossRef] [PubMed]
16. Somerville, W.R.C.; Law, A.D.; Rey, M.; Vogel, N.; Archer, A.J.; Buzza, D.M.A. Pattern formation in two-dimensional hard-core/soft-shell systems with variable soft shell profiles. *Soft Matter* **2020**, *16*, 3564. [CrossRef] [PubMed]
17. Ciarella, S.; Rey, M.; Harrer, J.; Holstein, N.; Ickler, M.; Löwen, H.; Vogel, N.; Janssen, L.M.C. Soft particles at liquid interfaces: From molecular particle architecture to collective phase behavior. *arXiv* **2020**, arXiv:2008.13695v1.
18. Almarza, N.G.; Pękalski, J.; Ciach, A. Periodic ordering of clusters and stripes in a two-dimensional lattice model. II. Results of Monte Carlo simulation. *J. Chem. Phys.* **2014**, *140*, 164708. [CrossRef] [PubMed]

© 2020 by the authors. Licensee MDPI, Basel, Switzerland. This article is an open access article distributed under the terms and conditions of the Creative Commons Attribution (CC BY) license (http://creativecommons.org/licenses/by/4.0/).

Article

Ultracold Bosons on a Regular Spherical Mesh

Santi Prestipino

Dipartimento di Scienze Matematiche ed Informatiche, Scienze Fisiche e Scienze della Terra, Università degli Studi di Messina, Viale F. Stagno d'Alcontres 31, 98166 Messina, Italy; sprestipino@unime.it

Received: 22 October 2020; Accepted: 11 November 2020; Published: 13 November 2020

Abstract: Here, the zero-temperature phase behavior of bosonic particles living on the nodes of a regular spherical mesh ("Platonic mesh") and interacting through an extended Bose-Hubbard Hamiltonian has been studied. Only the hard-core version of the model for two instances of Platonic mesh is considered here. Using the mean-field decoupling approximation, it is shown that the system may exist in various ground states, which can be regarded as analogs of gas, solid, supersolid, and superfluid. For one mesh, by comparing the theoretical results with the outcome of numerical diagonalization, I manage to uncover the signatures of diagonal and off-diagonal spatial orders in a finite quantum system.

Keywords: ultracold quantum gases; quantum phase transitions; decoupling approximation; spherical boundary conditions

1. Introduction

Gases of ultracold bosonic atoms loaded in an optical lattice provide the unique opportunity to study quantum many-body effects under controlled conditions [1,2]. To a very good approximation, the atoms can be described by a Bose–Hubbard (BH) Hamiltonian [3] whose parameters can be tuned by laser light [4,5]. By changing the configuration of the lasers, many lattice geometries can be explored, making optical lattices a powerful and versatile tool.

In a system at zero temperature ($T = 0$), thermal fluctuations are frozen out and quantum fluctuations prevail. These microscopic fluctuations can induce phase transitions in the ground state of a many-body system, driven by a non-thermal control parameter, such as chemical potential, magnetic field, or chemical composition. As a concrete example, consider a dilute gas of bosons at temperatures low enough that a Bose-Einstein condensate is formed. The condensate is described by a wave function consisting of every particle in the same state spread over the entire volume of the system, and typically exhibits superfluidity. An interesting situation appears when the condensate is subject to a lattice potential in which particles can move from one lattice site to the next only by tunneling. If the lattice potential is increased smoothly, the system remains in the condensed phase as long as the repulsion between atoms is small compared to the tunnel coupling (assuming, by the way, that the range of the repulsion is much smaller than the lattice spacing). In this regime, where the tunneling term dominates the Hamiltonian, a delocalized wave function still minimizes the total energy of the many-body system. When the strength of the repulsion becomes large compared to the tunnel coupling, the total energy is made minimum when each lattice site is occupied by the same number of atoms; as a result, phase coherence is lost and the system becomes insulating. The addition of a longer-range repulsion will make the phase behavior richer, with the possibility of a non-superfluid density wave and a supersolid ground state where crystalline order coexists with superfluid behavior (see, e.g., Ref. [6]). Experimentally, a way to prepare ultracold gases of long-range interacting bosons is to use atoms (such as chromium [7] or dysprosium [8]) and molecules having a large magnetic or electric dipole moment.

The usual BH model predicts a $T = 0$ phase transition from a superfluid phase to a Mott insulator phase as the ratio of the hopping matrix element between adjacent sites (t, in absolute terms) to the on-site interaction (U) is reduced. The overall number density of particles is controlled via a chemical-potential parameter μ. As μ grows at fixed t and U, the lattice becomes increasingly filled with particles, but this can only occur outside the Mott regions since the insulator phase is incompressible [3]. The BH model has been studied in many lattice geometries and with several techniques (mean-field theory [9–11], perturbation theory [12–16], and quantum Monte Carlo simulation [17,18], to name but the most commonly employed). When a further repulsion V is introduced between nearest-neighbor atoms ("extended BH model"), new phases may arise, in primis a supersolid phase [19–27]. Another variant of the BH model is hard-core bosons, where site occupancy is restricted to zero or one, corresponding to taking the $U \to \infty$ limit [6,28–32].

I hereafter present the results of yet another investigation of the extended BH model, now choosing a finite graph as hosting space for bosons. Even though clearcut phase transitions (i.e., thermodynamic singularities) cannot occur in a few-particle system, a convenient choice of boundary conditions may alleviate the difference with an infinite system, making the study of a finite quantum system valuable anyway. A practical solution is to use spherical boundary conditions (SBCs), which have often been exploited in the past to discourage long-range triangular ordering at high density [33–38]. On the other hand, SBCs make it possible to observe novel forms of ordering, viz. into regular polyhedral structures, that are simply unknown to Euclidean space—see Refs. [39–41]. An added value of a spherical mesh of points is the possibility to vary the site coordination while keeping the overall geometry strictly two-dimensional. Bosons confined in spherical (bubble) traps have been produced experimentally [42,43] and are going to be studied soon under microgravity conditions [44,45]. In the present study, the extended BH model is considered on a finite mesh of points homogeneously distributed on the unit sphere, i.e., coincident with the vertices of a regular polyhedron inscribed in the sphere (we may call it a Platonic mesh). Despite consisting of a finite number of nodes, a regular spherical mesh shares an important feature in common with an ordinary lattice, namely all sites are equivalent; for this reason, the phase behavior of particles living on a Platonic mesh would not deviate much from that of an infinite system (this intuition will be checked in a particular case).

The plan of this paper is the following. After introducing the models in Section 2, the choice of the underlying mesh is discussed in detail, giving priority to those regular grids where a subset of nodes form a mesh that is also regular. Then, in Section 3, a mean-field (MF) analysis of the ground state is carried out for all the models considered; in one case, the indications of theory are validated against the results of exact diagonalization (Section 4). From this comparison, we find the artifacts of the MF approximation as applied to a finite system. Finally, some concluding remarks are given in Section 5.

2. Particle Models on a Spherical Mesh

2.1. Classical Models

To illustrate the main idea, let it initially be considered a system of classical point particles defined on the sites of a cubic mesh stretched over the surface of the unit sphere. Each of the eight nodes of the mesh has three nearest neighbors (NN, at chord distance $r = 2/\sqrt{3}$) and three next-nearest neighbors (NNN, at chord distance $2\sqrt{2/3}$). For simplicity, assume that the occupancy n_i of site i can only be 0 or 1 ($i = 1, \ldots, 8$). Finally, choose the system Hamiltonian to be $H[n] = V \sum_{\langle i,j \rangle} n_i n_j$ with $V > 0$ (each NN pair in the sum is counted only once). The grand potential $\Omega(T, \mu)$ of this system is

$$\Omega = -\beta^{-1} \ln \Xi \quad \text{with} \quad \Xi = \sum_{\{n\}} e^{-\beta(H[n] - \mu \sum_i n_i)}, \tag{1}$$

where β is the inverse temperature. At $T = 0$, the formula is simpler:

$$\Omega = \min_{\{n\}} \left(H[n] - \mu \sum_i n_i \right). \tag{2}$$

Any of the points of absolute minimum in (2) represent the actual equilibrium state of the system for that μ. For particles living in a continuous space, minimization of $H - \mu N$ is carried out among a selection of crystalline states that are thought to be relevant based on symmetry considerations (see, e.g., Ref. [46]). In the present case, where the total number of microstates is small ($2^8 = 256$), the $T = 0$ grand potential can be determined exactly for each μ by a scrutiny of all possible energies. While the mesh is empty for $\mu < 0$ and completely filled with particles for $\mu > 3V$, in the interval from 0 to $3V$ only half of the nodes are occupied, and these fall at the vertices of a regular tetrahedron. This twofold-degenerate state with checkerboard order can be viewed as the finite-size counterpart of a crystalline phase.

The rationale behind the choice of a cubic mesh is now clear: by introducing a repulsion between occupied NN sites, we promote the occurrence of a Platonic "crystal", i.e., the regular tetrahedron ("CT model", see Figure 1 left). There is only one other possibility to obtain a non-frustrated crystal-like state at $T = 0$, which is using a regular dodecahedral mesh. We shall see that (i) by discouraging the occupancy of NN sites, a cubic "crystal" is stabilized for sufficiently small $\mu > 0$ ("DC model", Figure 1 center); (ii) if the repulsion is extended to embrace NNN sites too, then a tetrahedral "crystal" is stabilized in a range of positive μ values ("DT model", Figure 1 right).

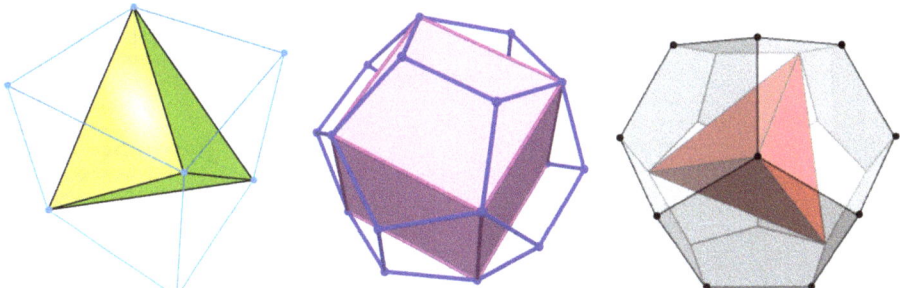

Figure 1. The two regular meshes considered in this work (the circumscribed sphere is not shown). The dots (some are hidden) are the sites/nodes of the mesh; their number is 8 for the cubic mesh and 20 for the regular dodecahedral mesh. (**Left**) Cubic mesh with a regular tetrahedron inside (CT model). (**Center**) Regular dodecahedral mesh with a cube inside (DC model). (**Right**) Regular dodecahedral mesh with a regular tetrahedron inside (DT model).

For a system of classical hard-core particles defined on a regular dodecahedral mesh, the total number of microstates is $2^{20} = 1,048,576$. By direct inspection, we see that the minimum-Ω "phase" for the DC model is the empty mesh for $\mu < 0$, a cube for $0 < \mu < 3V/2$ (a fivefold-degenerate state, since there are 5 ways to form a cube with the vertices of a regular dodecahedron), and the complement of a cube ("co-cube" in the following) for $3V/2 < \mu < 3V$; finally, for $\mu > 3V$ the mesh is completely filled. For the DT model, the mesh is empty for $\mu < 0$, filled with particles located at the vertices of a regular tetrahedron for $0 < \mu < 3V/2$ (a tenfold degenerate state), and completely filled for $\mu > 9V$. For μ between $3V/2$ and $9V$, a different ground state exists for each even number of occupied sites in the range 6 to 16, none of which corresponds to a simple geometric arrangement.

I have plotted in Figure 2 the evolution with temperature of a few properties of the CT and DC models. In addition to the grand potential Ω, the figure shows the total number of occupied sites (N), the total energy (E), and the order parameters for tetrahedral (S_t) and cubic order (S_c). The latter quantities are defined so as to discriminate the Platonic "phase" from the other $T = 0$ states. A proper order parameter should be insensitive to the orientation of the polyhedron, hence it can only depend on the relative angles between the vertices vector radii departing from the sphere center: for a regular tetrahedron all these angles are equal to $\alpha = \arccos\{-1/3\}$ (with $\sin \alpha = 2\sqrt{2}/3$), while they are $\pi - \alpha, \alpha$, and π for a cube. With the idea to penalize configurations that do not match the wanted structure, my choice of the OPs is the following:

$$S_t = \begin{cases} 1 - k_t \sum_{i<j} \left(\cos \theta_{ij} + \frac{1}{3}\right)^2 & , N = 4 \\ 0 & , N \neq 4 \end{cases} \quad (3)$$

and

$$S_c = \begin{cases} 1 - k_c \sum_{i<j} \sin \theta_{ij} \left(\sin \theta_{ij} - \frac{2}{3}\sqrt{2}\right)^2 & , N = 8 \\ 0 & , N \neq 8, \end{cases} \quad (4)$$

where the constants k_t and k_c are chosen, following the advice in Ref. [47], so that S vanishes for a random distribution of angles:

$$k_t = \frac{3}{8} \quad \text{and} \quad k_c = \frac{36}{7}\left(59\pi - 128\sqrt{2}\right)^{-1}. \quad (5)$$

Looking at Figure 2, it is clear that the only singularities are found at $T = 0$ (the system is finite) while all cusps and jumps are smoothened for $T > 0$.

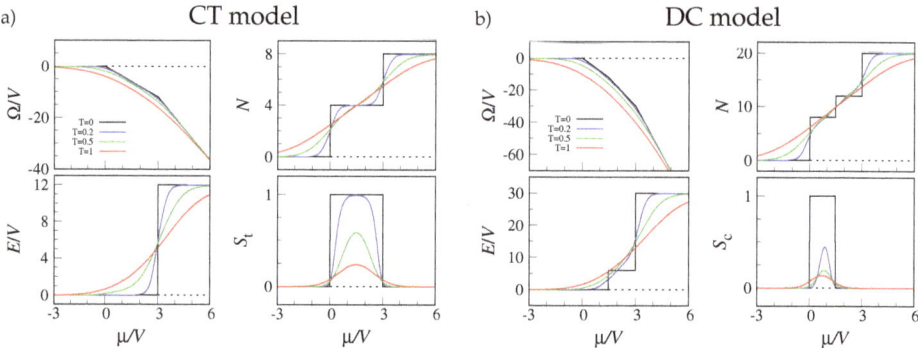

Figure 2. Thermal averages for the CT model (**a**) and the DC model (**b**), plotted as a function of μ for four values of T in units of V/k_B, where k_B is Boltzmann's constant (see legend in the **top left** panel). (**Top left**): Grand potential. (**Top right**): Total number of occupied sites. (**Bottom left**): Total energy. (**Bottom right**): Order parameter (see text).

It is worth considering how the $T = 0$ phase diagram of the CT and DC models gets modified when the occupancy of a node is allowed to take any value. For simplicity, as relevant $T = 0$ states only the "cluster crystals" [48–50] originated from the previously identified ground states are considered. Call $U > 0$ the on-site energy and V the NN repulsion. The grand Hamiltonian now reads $H[n] = U/2 \sum_i n_i(n_i - 1) + V \sum_{\langle i,j \rangle} n_i n_j - \mu \sum_i n_i$. For the CT model, the grand potential of a tetrahedral "cluster crystal" with n particles per site is $\Omega_n^{(4)} = 2n(n-1)U - 4n\mu$, while the grand potential of the "cluster crystal" with n particles per site is $\Omega_n^{(8)} = 4n(n-1)U + 12n^2V - 8n\mu$. For a given μ, the most stable "phase" corresponds to the minimum Ω. Up to $\mu = 0$ the stable "phase" is still the empty mesh. As μ grows further, the site occupancy increases monotonically within each of the families $\Omega_n^{(4)}$ and $\Omega_n^{(8)}$, but the exact sequence of stable "phases" depends on U/V. I show three cases in Figure 3. For $U = 5V$ (right panel), the behavior for small $\mu > 0$ recalls that found in the hard-core limit; however, as μ grows, each site becomes increasingly populated, and a whole sequence of cluster states is found. The opposite occurs for $U = 2V$ (left panel), where only the cluster states with checkerboard order are stabilized for $\mu > 0$. A curious situation occurs for $U = 3V$, where the competition between different cluster states becomes so stringent that "crystals" of the two families coexist at regular intervals along the μ axis (center panel).

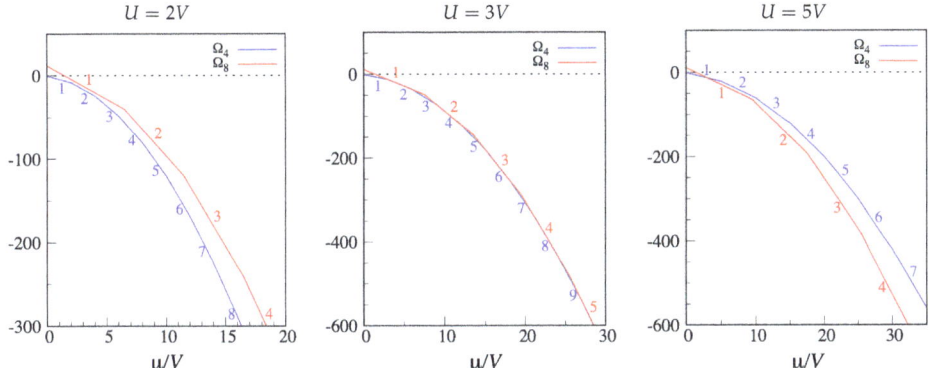

Figure 3. CT model with multiple occupancy allowed. The grand potential Ω (in V units) is plotted as a function of μ for two families of "cluster crystals", $\Omega_n^{(4)}$ (blue) and $\Omega_n^{(8)}$ (red), and for three values of U/V. Red and blue numbers are values of n.

As a matter of fact, hybrid ground states with features intermediate between those of the tetrahedral and cubic cluster states may also occur. For example, I have checked for $U = 5V$ that a microstate with two overlapping particles at the vertices of one tetrahedron and one particle at the vertices of the other tetrahedron is the most stable ground state in the μ interval from $8V$ to $11V$.

Moving to the DC model, we have three different families of "cluster phases": the cubic family with grand potential $\Omega_n^{(8)} = 4n(n-1)U - 8n\mu$; the co-cubic family with grand potential $\Omega_n^{(12)} = 6n(n-1)U + 6n^2V - 12n\mu$; and the family of those cluster states where every site is filled with the same number of particles, with grand potential $\Omega_n^{(20)} = 10n(n-1)U + 30n^2V - 20n\mu$. Without delving into details, the phase behavior as a function of U is similar to the CT model, with multiple coexistence between all three types of cluster crystals now occurring for $U = 2V$.

2.2. Quantum Models

The previous analysis has served to set the stage for the forthcoming study of the extended BH model on a regular spherical mesh; only the quantum versions of the CT and DC models (denoted QCT and QDC, respectively) are considered below.

The system is defined by the following grand Hamiltonian:

$$H = -t \sum_{\langle i,j \rangle} \left(a_i^\dagger a_j + a_j^\dagger a_i \right) + \frac{U}{2} \sum_i n_i(n_i - 1) + V \sum_{\langle i,j \rangle} n_i n_j - \mu \sum_i n_i. \tag{6}$$

In the present investigation, H describes a system of bosons on a spherical mesh of M sites, $a_i, a_i^\dagger$ are bosonic field operators ($i = 1, \ldots, M$), and $n_i = a_i^\dagger a_i$ is a number operator. Moreover, $t \geq 0$ is the hopping amplitude between NN sites, $U > 0$ is the on-site repulsion, and $V > 0$ is the strength of the NN repulsion. In the hard-core limit $U \to \infty$, the site occupancy is restricted to zero or one and the U term in (6) vanishes; it is only this limit that is treated hereafter.

In the BH model on a standard lattice, at $T = 0$ we observe an insulator–superfluid transition with increasing t for every positive μ. The addition of V may stabilize (depending on the lattice) a density wave at low t, as well as a supersolid phase at the boundary between the insulator and superfluid phases. These features are also present in the hard-core limit, as reported, e.g., in Refs. [30,32].

3. MF Investigation

MF theory is the method of choice when a new many-body problem is attacked; it has been frequently applied for continuous quantum systems as well, as an effective means to identify the ground states and quantum transitions between them (cf. Refs. [51–53]). Various versions of MF theory exist in discrete space; here the so-called decoupling approximation [11,29,32] is employed, which gives the advantage of a fully analytic treatment of the problem. Clearly, the accuracy of MF theory would be questionable when applied for a finite quantum system, even one lacking a boundary surface. This will make urgent an assessment of MF theory against exact results, which however is delayed until the next Section.

In the decoupling approximation, the two-site terms in the Hamiltonian (6) are linearized so that the eigenvalue problem becomes effectively one-site. This is accomplished by first writing the exact identity

$$a_i^\dagger a_j = a_i^\dagger \langle a_j \rangle + \langle a_i^\dagger \rangle a_j - \langle a_i^\dagger \rangle \langle a_j \rangle + \delta a_i^\dagger \delta a_j \tag{7}$$

with $\delta a_i = a_i - \langle a_i \rangle$ (for $i = 1, \ldots, M$), and then neglecting the last term. Similarly,

$$n_i n_j \approx n_i \langle n_j \rangle + \langle n_i \rangle n_j - \langle n_i \rangle \langle n_j \rangle . \tag{8}$$

The averages $\langle a_i \rangle \equiv \phi_i$ and $\langle n_i \rangle \equiv \rho_i$ (either ground-state averages or thermal averages) are to be determined self-consistently; ϕ_i and ρ_i (with $0 \leq \rho_i \leq 1$) represent the superfluid order parameter and the local density for site i, respectively. For hard-core bosons, we readily obtain $H \approx H_{\mathrm{MF}}$ with

$$H_{\mathrm{MF}} = -t \sum_i \left(F_i a_i^\dagger + F_i^* a_i - F_i \phi_i^* \right) + \frac{V}{2} \sum_i (2R_i n_i - R_i \rho_i) - \mu \sum_i n_i . \tag{9}$$

In Equation (9), $F_i = \sum_{j \in \mathrm{NN}_i} \phi_j$ and $R_i = \sum_{j \in \mathrm{NN}_i} \rho_j$ are sums over the nearest neighbors of the i-th site. For a bipartite mesh, sites are either of type A or B, hence the unknown parameters are four, namely $\phi_A, \phi_B, \rho_A, \rho_B$. The simplification is obvious: rather than working on a Fock space of dimensionality 2^M, for a bipartite mesh the basis states are just four, namely $|0,0\rangle, |0,1\rangle, |1,0\rangle,$ and $|1,1\rangle$, corresponding to the possible occupancies of a pair of A and B sites.

3.1. QCT Model

For the QCT model, the mesh is bipartite and formed by two tetrahedral sub-meshes with four points each. A point of grid A has three neighbors, all belonging to grid B, and *vice versa*. Hence,

$$F_A = 3\phi_B, \quad F_B = 3\phi_A, \quad R_A = 3\rho_B, \quad \text{and} \quad R_B = 3\rho_A . \tag{10}$$

Using a subscript A (B) for operators relative to a single A (B) site, the MF Hamiltonian reads:

$$H_{\mathrm{MF}} = E_0 - 12t \left(\phi_B a_A^\dagger + \phi_B^* a_A + \phi_A a_B^\dagger + \phi_A^* a_B \right) + 4(3V\rho_B - \mu) n_A + 4(3V\rho_A - \mu) n_B \tag{11}$$

with

$$E_0 = 12t(\phi_A \phi_B^* + \phi_A^* \phi_B) - 12V\rho_A \rho_B . \tag{12}$$

On the 4-vector basis $\mathcal{B} = \{|0,0\rangle, |0,1\rangle, |1,0\rangle, |1,1\rangle\}$, the Hamiltonian (11) is represented by a 4×4 Hermitian matrix:

$$H_{\mathrm{MF}} = \begin{pmatrix} E_0 & -12t\phi_A^* & -12t\phi_B^* & 0 \\ -12t\phi_A & E_0 + 4(3V\rho_A - \mu) & 0 & -12t\phi_B^* \\ -12t\phi_B & 0 & E_0 + 4(3V\rho_B - \mu) & -12t\phi_A^* \\ 0 & -12t\phi_B & -12t\phi_A & E_0 + 12V(\rho_A + \rho_B) - 8\mu \end{pmatrix} . \tag{13}$$

Non-superfluid phases have $\phi_A = \phi_B = 0$. In this case, the matrix becomes diagonal, meaning that the basis vectors are all (grand-)energy eigenstates. From the self-consistency equations $\rho_A = \langle n_A \rangle$ and $\rho_B = \langle n_B \rangle$, it readily follows that $\rho_A = \rho_B = 0$ for $|0,0\rangle$ (empty mesh), $\rho_A = 0$ and $\rho_B = 1$ for $|0,1\rangle$ (tetrahedral-B "crystal"), $\rho_A = 1$ and $\rho_B = 0$ for $|1,0\rangle$ (tetrahedral-A "crystal"), and $\rho_A = \rho_B = 1$ for $|1,1\rangle$ (fully occupied mesh). The eigenvalue gives the grand potential Ω, which is zero for the empty mesh, -4μ for the twofold-degenerate tetrahedral "crystal", and $12V - 8\mu$ for the filled mesh. These grand potentials are exactly the same as in the CT model.

Superfluid and supersolid "phases" have non-zero, possibly distinct, complex values of ϕ_A and ϕ_B. For sure, ϕ_A and ϕ_B have equal phases since only the magnitude of the order parameter can be spatially modulated; hence, the arbitrary phase can be taken as zero. Without loss of generality, we may assume ϕ_A and ϕ_B to be positive quantities. We shall see in the next section that the exact eigenstates of H for $t > 0$ do not distinguish between A and B; hence, our search can be restricted to homogeneous (superfluid) solutions: $\phi_A = \phi_B = \phi$ and $\rho_A = \rho_B = \rho$. We are thus led to the following characteristic equation (with $E_0 = 24t\phi^2 - 12V\rho^2$):

$$\begin{vmatrix} E_0 - \lambda & -12t\phi & -12t\phi & 0 \\ -12t\phi & E_0 + 4(3V\rho - \mu) - \lambda & 0 & -12t\phi \\ -12t\phi & 0 & E_0 + 4(3V\rho - \mu) - \lambda & -12t\phi \\ 0 & -12t\phi & -12t\phi & E_0 + 8(3V\rho - \mu) - \lambda \end{vmatrix} = 0. \quad (14)$$

Doing the simplifications, we arrive at

$$(\lambda - b)^2 \left[(\lambda - a)(\lambda - 2b + a) - 4u^2 \right] = 0 \quad (15)$$

with $a = E_0$, $b = E_0 + 4(3V\rho - \mu)$, and $u = -12t\phi$. The minimum root of (15) is

$$E = b - \sqrt{(a-b)^2 + 4u^2} = 4\left(6t\phi^2 - 3V\rho^2 + 3V\rho - \mu - \sqrt{(3V\rho - \mu)^2 + 36t^2\phi^2} \right). \quad (16)$$

A real eigenvector of E is

$$|\psi_E\rangle = \frac{E - 2b + a}{2u} |0,0\rangle + |0,1\rangle + |1,0\rangle + \frac{E - a}{2u} |1,1\rangle, \quad (17)$$

or, in explicit terms,

$$|\psi_E\rangle = \left(\frac{3V\rho - \mu + \sqrt{(3V\rho - \mu)^2 + 36t^2\phi^2}}{6t\phi}, 1, 1, \frac{-(3V\rho - \mu) + \sqrt{(3V\rho - \mu)^2 + 36t^2\phi^2}}{6t\phi} \right). \quad (18)$$

Now imposing the conditions

$$\rho_{A,B} = \frac{\langle \psi_E | n_{A,B} | \psi_E \rangle}{\langle \psi_E | \psi_E \rangle} \quad \text{and} \quad \phi_{A,B} = \frac{\langle \psi_E | a_{A,B} | \psi_E \rangle}{\langle \psi_E | \psi_E \rangle}, \quad (19)$$

we arrive at the two coupled equations

$$1 - 2\rho = \frac{3V\rho - \mu}{\sqrt{(3V\rho - \mu)^2 + 36t^2\phi^2}} \quad \text{and} \quad 1 = \frac{3t}{\sqrt{(3V\rho - \mu)^2 + 36t^2\phi^2}}, \quad (20)$$

which are easily solved to give

$$\rho = \frac{\mu + 3t}{3V + 6t} \quad \text{and} \quad \phi = \frac{\sqrt{(\mu + 3t)(3V + 3t - \mu)}}{3V + 6t}. \quad (21)$$

The above ϕ solution only exists provided that $-3t < \mu < 3V + 3t$, which is a necessary condition for the existence of the superfluid. Plugging these ρ and ϕ in Equation (16), we finally obtain the superfluid grand potential:

$$\Omega_{SF} = -\frac{4(\mu + 3t)^2}{3V + 6t}. \tag{22}$$

This outcome can also be obtained from the general relation (see Equations (6)–(8))

$$\langle H_{MF} \rangle = -2t \sum_{\langle i,j \rangle} \phi_i^* \phi_j + V \sum_{\langle i,j \rangle} \rho_i \rho_j - \mu \sum_i \rho_i. \tag{23}$$

For the superfluid, we have $\phi_i = \phi$ and $\rho_i = \rho$, with ϕ and ρ given by Equation (21); hence,

$$\langle H_{MF} \rangle = 4 \left(-6t\phi^2 + 3V\rho^2 - 2\mu\rho \right) = -\frac{4(\mu + 3t)^2}{3V + 6t}. \tag{24}$$

By comparing the grand potentials of all the "phases", we arrive at the ground-state diagram in Figure 4. Along the straight lines $\mu = -3t$ and $\mu = 3V + 3t$, separating the superfluid from the insulator "phases" at small and large μ, the μ derivative of the grand potential ($= -\langle N \rangle$) is continuous. Instead, along the line

$$\mu = \frac{3V}{2} \pm \frac{3}{2}\sqrt{V^2 - 4t^2}, \tag{25}$$

separating the "crystalline" lobe from the superfluid, the average number of occupied sites shows a jump discontinuity. Only at the lobe vertex $(V/2, 3V/2)$ the tetrahedral-superfluid transition is continuous. We will see in Section 4 to what extent the indications of MF theory are accurate, considering that the cubic mesh consists of 8 sites only.

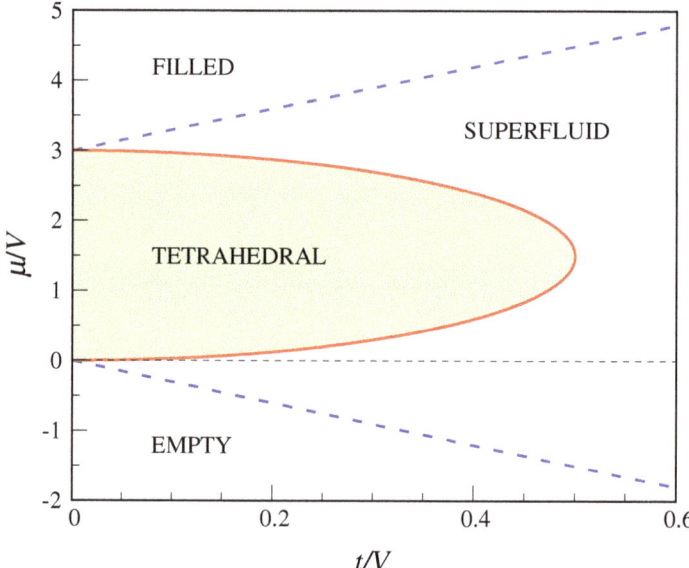

Figure 4. Phase diagram of the QCT model according to the decoupling approximation (see text). The blue dashed lines are second-order transition lines, while the red full line is a first-order line (only at the lobe vertex the transition is continuous).

3.2. QDC Model

The decoupling approximation for the QDC model works similarly as for the QCT model. Again, the starting point is Equation (9) and the mesh is partitioned into a cube (A, 8 sites) and a co-cube (B, 12 sites). We have:

$$F_A = 3\phi_B, \quad F_B = 2\phi_A + \phi_B, \quad R_A = 3\rho_B, \quad \text{and} \quad R_B = 2\rho_A + \rho_B. \tag{26}$$

The MF Hamiltonian then reads:

$$\begin{aligned} H_{\text{MF}} &= E_0 - 24t(\phi_B a_A^\dagger + \phi_B^* a_A) - 12t\left((2\phi_A + \phi_B)a_B^\dagger + (2\phi_A^* + \phi_B^*)a_B\right) \\ &\quad + 4(6V\rho_B - 2\mu)n_A + 4(6V\rho_A + 3V\rho_B - 3\mu)n_B \end{aligned} \tag{27}$$

with

$$E_0 = 24t(\phi_A \phi_B^* + \phi_A^* \phi_B) + 12t|\phi_B|^2 - 24V\rho_A\rho_B - 6V\rho_B^2. \tag{28}$$

On the $\mathcal{B}$ basis, the Hamiltonian (27) is represented by the matrix:

$$H_{\text{MF}} = \begin{pmatrix} E_0 & -12t(2\phi_A^* + \phi_B^*) & -24t\phi_B^* & 0 \\ -12t(2\phi_A + \phi_B) & E_0 + 4(6V\rho_A + 3V\rho_B - 3\mu) & 0 & -24t\phi_B^* \\ -24t\phi_B & 0 & E_0 + 4(6V\rho_B - 2\mu) & -12t(2\phi_A^* + \phi_B^*) \\ 0 & -24t\phi_B & -12t(2\phi_A + \phi_B) & E_0 + 4(6V\rho_A + 9V\rho_B - 5\mu) \end{pmatrix}. \tag{29}$$

For phases with $\phi_A = \phi_B = 0$, the matrix is diagonal and, like in the QCT model, the basis vectors are all (grand-)energy eigenstates. From the self-consistency conditions $\rho_A = \langle n_A \rangle$ and $\rho_B = \langle n_B \rangle$ it follows that $\rho_A = \rho_B = 0$ for $|0,0\rangle$ (empty mesh), $\rho_A = 0$ and $\rho_B = 1$ for $|0,1\rangle$ (co-cubic "crystal"), $\rho_A = 1$ and $\rho_B = 0$ for $|1,0\rangle$ (cubic "crystal"), and $\rho_A = \rho_B = 1$ for $|1,1\rangle$ (filled mesh). The grand potential Ω is zero for the vacuum, -8μ for the cubic "crystal", $6V - 12\mu$ for the co-cubic "crystal", and $30V - 20\mu$ for the filled mesh. These values are the same as for the DC model, hence the same sequence of "phases" as a function of μ is observed in the QDC model for $t = 0$.

Moving to phases with $\phi_A, \phi_B \neq 0$, I first consider the possibility of a superfluid ($\phi_A = \phi_B = \phi > 0$ and $\rho_A = \rho_B = \rho$). In this case, the characteristic equation takes the form:

$$\begin{vmatrix} a - \lambda & u & v & 0 \\ u & b - \lambda & 0 & v \\ v & 0 & c - \lambda & u \\ 0 & v & u & d - \lambda \end{vmatrix} = 0 \tag{30}$$

with $E_0 = 60t\phi^2 - 30V\rho^2$, $a = E_0$, $b = E_0 + 12(3V\rho - \mu)$, $c = E_0 + 8(3V\rho - \mu)$, $d = E_0 + 20(3V\rho - \mu)$, $u = -36t\phi$, $v = -24t\phi$. Observing that $a + d = b + c \equiv s$, the minimum eigenvalue of H_{MF} turns out to be (see Appendix A):

$$\begin{aligned} E &= \frac{1}{2}\left(s - \sqrt{(a-b)^2 + 4u^2} - \sqrt{(a+b-s)^2 + 4v^2}\right) \\ &= 10\left(6t\phi^2 - 3V\rho^2 + 3V\rho - \mu - \sqrt{(3V\rho - \mu)^2 + 36t^2\phi^2}\right). \end{aligned} \tag{31}$$

Notice that this energy is exactly 5/2 times that of the QCT model (see Equation (16)).

Using $p = (b - c)/2 = (d - a)/10 = 2(3V\rho - \mu)$ and $q = u/3 = v/2 = -12t\phi$, the coordinates (x, y, z, w) of an eigenvector $|\psi_E\rangle$ of E satisfy the following linear system:

$$\begin{cases} -5px + 3qy + 2qz = -5\sqrt{p^2+q^2}\,x \\ 3qx + py + 2qw = -5\sqrt{p^2+q^2}\,y \\ 2qx - pz + 3qw = -5\sqrt{p^2+q^2}\,z \\ 2qy + 3qz + 5pw = -5\sqrt{p^2+q^2}\,w, \end{cases} \qquad (32)$$

implying that $|\psi_E\rangle$ is also an eigenvector of the simpler matrix

$$\begin{pmatrix} -5p & 3q & 2q & 0 \\ 3q & p & 0 & 2q \\ 2q & 0 & -p & 3q \\ 0 & 2q & 3q & 5p \end{pmatrix} \qquad (33)$$

with eigenvalue $-5\sqrt{p^2+q^2}$. For $t > 0$, one such vector is

$$|\psi_E\rangle = \left(-\frac{p+\sqrt{p^2+q^2}}{q}, 1, 1, \frac{p-\sqrt{p^2+q^2}}{q} \right), \qquad (34)$$

or

$$|\psi_E\rangle = \left(\frac{3V\rho - \mu + \sqrt{(3V\rho-\mu)^2 + 36t^2\phi^2}}{6t\phi}, 1, 1, \frac{-(3V\rho-\mu) + \sqrt{(3V\rho-\mu)^2 + 36t^2\phi^2}}{6t\phi} \right). \qquad (35)$$

This is identical to the state (18) describing the superfluid in the QCT model. The reason of this equivalence is that, in both cubic and dodecahedral meshes, every site has three neighbors; hence, in the superfluid, the quantities F_A, F_B, R_A, R_B are the same. No surprise, then, if also the expressions of ρ and ϕ turn out to be equal for the two models (the consistency equations are identical). As already commented, the energy of the QDC superfluid is instead a factor $5/2$ larger (in absolute terms) than in the QCT model:

$$\Omega_{SF} = -\frac{10(\mu + 3t)^2}{3V + 6t}. \qquad (36)$$

The main novelty with respect to the QCT model is the existence of a stable supersolid "phase" in the QDC model. To seek for MF solutions having the character of a supersolid, I have first derived the exact equations for the MF parameters $\rho_A, \rho_B, \phi_A, \phi_B$, without *a priori* assuming them to be equal in pairs (this is done in Appendix B). Then, I have solved these equations numerically so as to find the region where the supersolid is more stable than the other phases. This task is made simpler by noting that, thanks to a symmetry property of the equations, from a supersolid solution with $\mu > 3V/2$ it is possible to obtain another solution with $\mu < 3V/2$. Indeed, it easily follows from Equation (A15) that, if $(\rho_A, \rho_B, \phi_A, \phi_B)$ is a solution for a certain μ, then $(1-\rho_A, 1-\rho_B, \phi_A, \phi_B)$ is a solution for $3V - \mu$. Moreover, for $\mu = 3V/2$ the two solutions share the same grand potential. It turns out that the (t, μ) region where the supersolid is stable is symmetric about the $\mu = 3V/2$ axis, lying across the boundary between the solid "phases" and the superfluid (see Figure 5). To all evidence, the boundary between the supersolid and the superfluid lies at $t = 1/3$. In fact, the supersolid consists of two distinct "phases", SS1 below $\mu = 3V/2$ (where $\rho_B < 1/2 < \rho_A$) and SS2 above $\mu = 3V/2$ (where $\rho_A < 1/2 < \rho_B$), with $\phi_A < \phi_B < 1/2$ in both. In SS1, cubic sites are more occupied, on average, than co-cubic sites, whereas the opposite occurs in SS2. The two supersolids coexist for $\mu = 3V/2$, for all t in the range 0.25 to 1/3. Hard-core bosons on the triangular lattice have a similar phase diagram [32], but in that case the supersolid extends down to $t = 0$ and the solid lobes do not overlap each other.

The full $T = 0$ phase diagram of the QDC model according to the decoupling approximation is reported in Figure 5. Along the straight lines $\mu = -3t$ and $\mu = 3V + 3t$, the μ derivative of the grand potential is continuous. Instead, along the curves

$$\mu = \frac{-3t + 6V \pm 6\sqrt{V^2 - Vt - 6t^2}}{5} \quad \text{and} \quad \mu = \frac{3t + 9V \pm 6\sqrt{V^2 - Vt - 6t^2}}{5}, \tag{37}$$

which respectively separate the cubic and co-cubic regions from the superfluid region, the total number of particles jumps discontinuously. Only at the lobe vertices $(V/3, V)$ and $(V/3, 2V)$ the two "crystal"—superfluid transitions are continuous. The line $\mu = 3V/2$ separating the two "crystalline" regions ends at the point $(3V/10, 3V/2)$ where the two boundary curves (37) cross each other. However, this triple point is only metastable since, in the whole region bounded by the blue circuit of Figure 5, the stable phase is actually supersolid.

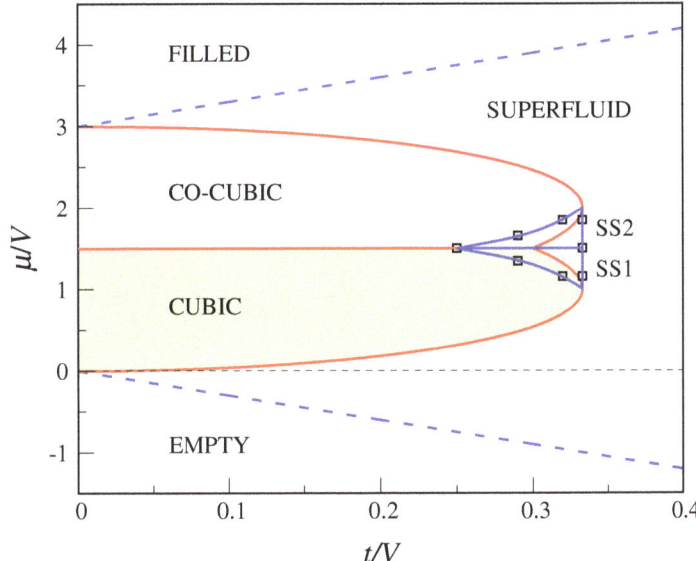

Figure 5. Phase diagram of the QDC model according to the decoupling approximation (see text). The blue dashed lines are second-order transition lines, while red full lines are first-order lines (only at the lobe vertices the transition is continuous). Inside the blue circuit, the stable phase is supersolid (SS1 below $\mu = 3V/2$; SS2 above $\mu = 3V/2$). The full lines through the data points (open black squares) are a guide for the eye.

4. Assessment of MF Theory

Let us again reconsider the QCT model at $T = 0$. Due to the relatively small dimensionality of its Hilbert space (=256), the model can be solved numerically, determining (among others) the exact ground state $|g\rangle$ and its eigenvalue (i.e., the grand potential) as a function of t and μ. This is obtained by representing the grand Hamiltonian H on the Fock basis $\{|x_1, x_2, \ldots, x_8\rangle\}$ (with $x_i = 0$ or 1 for all i) and diagonalizing the ensuing matrix. An extensive mapping of a few quantum averages enables us to clarify the nature of the "phases" present.

As usual, the sites of the mesh are classified according to what tetrahedral sub-mesh, A or B, they belong to. Then, I compute the average occupancies of A and B sites (corresponding to the MF parameters ρ_A and ρ_B, respectively), the average values of a_A and a_B (corresponding to the MF parameters ϕ_A and ϕ_B, respectively), and the superfluid density ρ_{SF} (see, e.g., Ref. [13,22]). In the present case, the latter quantity reads:

$$\rho_{\text{SF}} \equiv \frac{1}{8} \left\langle g \left| \widetilde{a}_0^\dagger \widetilde{a}_0 \right| g \right\rangle = \frac{1}{64} \sum_{i,j=1}^{8} \left\langle g \left| a_i^\dagger a_j \right| g \right\rangle = \frac{1}{8} \sum_{j=1}^{8} \left\langle g \left| a_1^\dagger a_j \right| g \right\rangle, \quad (38)$$

where $\widetilde{a}_0 = (1/\sqrt{8}) \sum_{i=1}^{8} a_i$ is the zero-momentum field operator. Notice that, in a large lattice of M sites, $\langle \widetilde{a}_0^\dagger \widetilde{a}_0 \rangle = N_0$ is the average number of condensate particles, hence $\rho_{\text{SF}} = N_0/M$ is indeed the condensate density.

As far as the occupancies are concerned, exact diagonalization shows that they are always equal for an A and a B site, with the only exception of $t = 0$ and $0 < \mu < 3V$ where the occupancies are as in MF theory (i.e., either $\langle n_A \rangle = 1$ and $\langle n_B \rangle = 0$ or vice versa). In particular, for $t > 0$ the equivalence $\langle n_A \rangle = \langle n_B \rangle$ holds in the whole (t, μ) region pertaining, according to MF theory, to the tetrahedral "phase". Hence, no spontaneous symmetry breaking does really occur in the QCT model, except for $t = 0$. Indeed, for $t > 0$ the Fourier coefficients of $|g\rangle$ relative to any pair of Fock states equal by A-B inversion are the same. I show in Figure 6 the average site occupancy in the whole space of parameters. We see that the (t, μ) plane is divided in zones where the site occupancy, which overall grows with μ, takes the same constant value. The possible values are $k/8$, for $k = 0, 1, \ldots, 8$. This outcome is not entirely unexpected considering that the Hamiltonian commutes with the N operator, implying that the ground state (which is non-degenerate for $t > 0$) should also be an eigenstate of N. Like in MF theory, the occupancy is zero below $\mu = -3t$ and 1 above $\mu = 3V + 3t$; in a whole region around $\mu = 3V/2$ the occupancy is 0.5 for all t, with no abrupt transition from "crystal" to superfluid values.

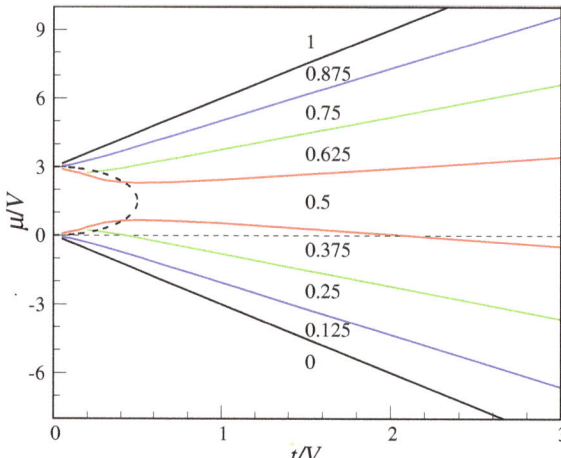

Figure 6. Exact average site occupancy in the QCT model (full symmetry occurs between the A and B sub-meshes). The plotted numbers are the occupancies in the whole regions delimited by the colored lines. These values reflect a perfect particle–hole symmetry around $\mu = 3V/2$. The MF boundary between the tetrahedral and superfluid "phases" is also shown for comparison (dashed line).

Another difference with MF theory concerns the ground-state averages of a_A and a_B: these are identically zero for all t and μ values, which may seem in stark contrast to the behavior of the MF parameter ϕ. In fact, it is not these averages that should be monitored in an exact treatment but rather the ρ_{SF} quantity defined at Equation (38). In the top panel of Figure 7, I have reported the ρ_{SF} values computed along a few iso-t lines; each jump discontinuity of ρ_{SF} occurs in coincidence with one of the site occupancy. For comparison, in the bottom panel of Figure 7, I show the MF values of ϕ^2 along the same lines. We see a clear correlation between the two behaviors, which demonstrates the existence of signatures of superfluidity in a finite quantum system. However, intriguing differences also exist: (i) first observe that $\rho_{SF} = 0.125$ in the filled mesh, which is an oddity for a perfectly insulating state. In fact, 0.125 is $1/M$ for $M = 8$, suggesting that this is an artifact of the finite system size. (ii) Another finite-size effect is the non-zero values of ρ_{SF} in the purported tetrahedral region. However, ρ_{SF} is here sufficiently small so that its peaks in the μ gaps between the tetrahedral and insulator regions remain well visible—in MF theory these gaps fall in the superfluid region.

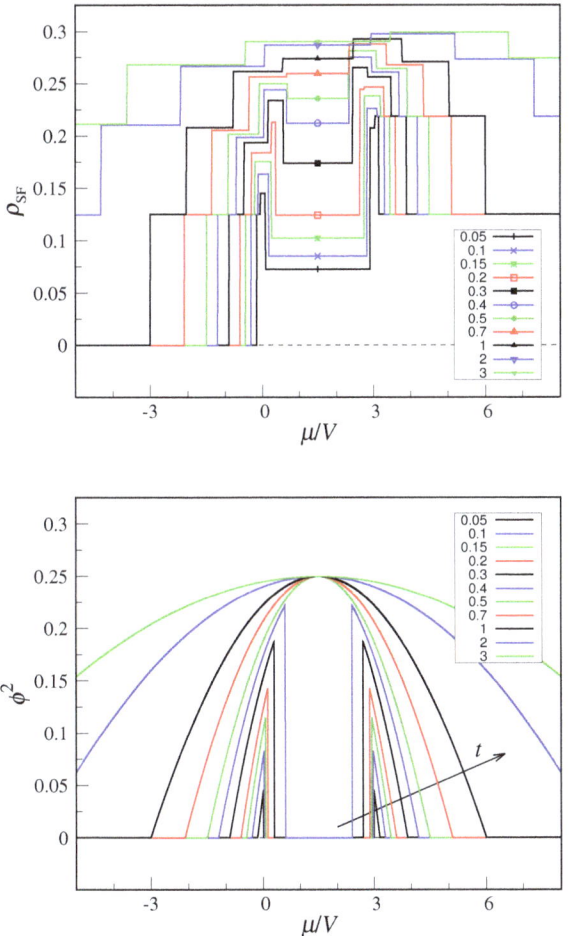

Figure 7. Superfluid density in the QCT model: comparison between exact diagonalization (**top panel**) and decoupling approximation (**bottom panel**). The superfluid density is plotted as a function of μ for a number of t values in units of V, increasing from bottom to top (see legends).

Finally, in Figure 8, I plot the QCT grand potential as a function of μ for a number of t values in the range 0 to 0.5. In the MF curves, cusp singularities are associated with the crossing of first-order transition lines. We see that MF theory systematically overestimates exact values, the more so the larger t is.

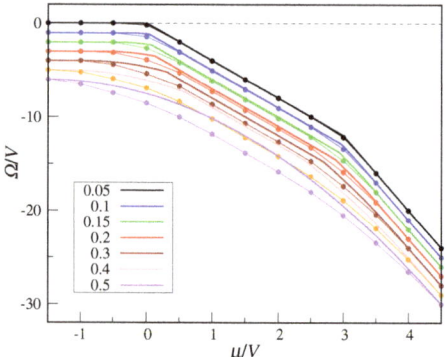

Figure 8. Grand potential of the QCT model as a function of μ, for a number of t values (from top to bottom, $t = 0.05, 0.1, 0.15, 0.2, 0.3, 0.4, 0.5$). MF data (thick full lines) are compared with exact values (full dots). To improve figure readability, the dots are joined by thin straight-line segments and the data are displaced vertically by 1 with respect to one another, starting from $t = 0.1$.

5. Conclusions

I have worked out the zero-temperature phase diagram of two systems of hard-core bosons defined on the nodes of a regular spherical mesh. The interaction is of Bose–Hubbard type, with a further repulsion between neighboring particles. Choosing a suitable mesh, bosons are pushed to form a Platonic "crystal" in a range of chemical potentials. In the QCT model, the mesh is cubic and a tetrahedral "crystal" is formed; in the QDC model, the mesh is dodecahedral and a cubic "crystal" is formed instead.

Using a mean-field approximation, I have obtained fully analytic results for the thermodynamic properties of the two models. In addition to a number of insulating phases, both systems also exhibit a superfluid ground state. In the QDC model, triple coexistence between two solids and the superfluid is superseded by the occurrence of a more stable supersolid phase. Clearly, while the predictions of mean-field theory are generally accurate for an infinite Bose–Hubbard system, deviations will unavoidably be observed in a small system, where no true singularity can occur. However, discrepancies are probably less strong for bosons on a regular spherical mesh, which has equivalent sites and is devoid of natural boundaries.

To check this expectation for the QCT model, I have diagonalized its Hamiltonian exactly, finding the ground state and a number of ground-state averages. Overall, the exact $T = 0$ behavior of the finite system does not depart much from theory, though obviously phase-transition lines are now reduced to simple crossovers; however, clear marks of superfluidity have been detected in the same region of parameters where MF theory predicts it to occur. In conclusion, I hope that the present work may stimulate research aimed at the realization of a new experimental platform for ultracold bosons that is akin to a regular spherical mesh.

Funding: This research received no external funding.

Conflicts of Interest: The author declares no conflict of interest.

Appendix A. On the Solutions to an Algebraic Equation

In this Appendix, I show how to determine the eigenvalues of a matrix like (29), namely the four real solutions λ to the characteristic Equation (30) with $a + d = b + c \equiv s$.

Upon doing the algebra, the characteristic equation turns out to be:

$$(\lambda - a)(\lambda - b)(\lambda - s + a)(\lambda - s + b) - u^2\left[(\lambda - a)(\lambda - b) + (\lambda - s + a)(\lambda - s + b)\right]$$
$$-v^2\left[(\lambda - a)(\lambda - s + b) + (\lambda - s + a)(\lambda - b)\right] + (u^2 - v^2)^2 = 0. \quad \text{(A1)}$$

Let us now reshuffle the lhs of (A1) term by term:

(i) $(\lambda - a)(\lambda - b)(\lambda - s + a)(\lambda - s + b)$
$$= \frac{1}{16}\left[(2\lambda - s)^2 - (2a - s)^2\right]\left[(2\lambda - s)^2 - (2b - s)^2\right]$$
$$= \frac{1}{16}\left\{(2\lambda - s)^4 - \left[(2a - s)^2 + (2b - s)^2\right](2\lambda - s)^2 + (2a - s)^2(2b - s)^2\right\} \quad \text{(A2)}$$
$$= \frac{1}{16}\left\{(2\lambda - s)^4 - 2\left[(a - b)^2 + (a + b - s)^2\right](2\lambda - s)^2 + \left[(a - b)^2 - (a + b - s)^2\right]^2\right\};$$

(ii) $-u^2\left[(\lambda - a)(\lambda - b) + (\lambda - s + a)(\lambda - s + b)\right] = -\frac{1}{2}u^2\left[(2\lambda - s)^2 + (a + b - s)^2 - (a - b)^2\right];$ (A3)

(iii) $-v^2\left[(\lambda - a)(\lambda - s + b) + (\lambda - s + a)(\lambda - b)\right] = -\frac{1}{2}v^2\left[(2\lambda - s)^2 + (a - b)^2 - (a + b - s)^2\right].$ (A4)

Plugging Equations (A2)–(A4) in (A1), the latter equation becomes:

$$(2\lambda - s)^4 - 2\left[(a - b)^2 + (a + b - s)^2 + 4(u^2 + v^2)\right](2\lambda - s)^2 + \left[(a - b)^2 - (a + b - s)^2\right]^2$$
$$+ 8(u^2 - v^2)\left[(a - b)^2 - (a + b - s)^2\right] + 16(u^2 - v^2)^2 = 0 \quad \text{(A5)}$$

or equivalently

$$(2\lambda - s)^4 - 2\left[(a - b)^2 + (a + b - s)^2 + 4(u^2 + v^2)\right](2\lambda - s)^2$$
$$+ \left[(a - b)^2 - (a + b - s)^2 + 4(u^2 - v^2)\right]^2 = 0. \quad \text{(A6)}$$

Upon defining the two quantities $X = (a - b)^2 + 4u^2$ and $Y = (a + b - s)^2 + 4v^2$, the lhs of Equation (A6) can be factorized as follows:

$$(2\lambda - s)^4 - 2(X + Y)(2\lambda - s)^2 + (X - Y)^2$$
$$= \left[(2\lambda - s)^2 - (X + Y)\right]^2 - 4XY$$
$$= \left[(2\lambda - s)^2 - X - Y + 2\sqrt{X}\sqrt{Y}\right]\left[(2\lambda - s)^2 - X - Y - 2\sqrt{X}\sqrt{Y}\right]$$
$$= \left[(2\lambda - s)^2 - \left(\sqrt{X} - \sqrt{Y}\right)^2\right]\left[(2\lambda - s)^2 - \left(\sqrt{X} + \sqrt{Y}\right)^2\right]. \quad \text{(A7)}$$

Hence, two independent second-order equations are obtained from (A6):

$$(2\lambda - s)^2 = \left(\sqrt{X} - \sqrt{Y}\right)^2 \quad \text{and} \quad (2\lambda - s)^2 = \left(\sqrt{X} + \sqrt{Y}\right)^2, \quad \text{(A8)}$$

whose solutions are the searched eigenvalues:

$$\lambda_1 = \frac{1}{2}\left(s - \sqrt{X} + \sqrt{Y}\right); \quad \lambda_2 = \frac{1}{2}\left(s + \sqrt{X} - \sqrt{Y}\right);$$
$$\lambda_3 = \frac{1}{2}\left(s - \sqrt{X} - \sqrt{Y}\right); \quad \lambda_4 = \frac{1}{2}\left(s + \sqrt{X} + \sqrt{Y}\right). \quad \text{(A9)}$$

The minimum eigenvalue is solution no. 3 above.

Appendix B. Self-Consistency Conditions in the QDC Model

I here derive the general equations obeyed by MF parameters in the QDC model. These are obtained by first determining the leading eigenvector of the matrix representing the MF grand Hamiltonian on $\mathcal{B}$. Then, the self-consistency conditions are written in the same way as Equation (19).

The MF Hamiltonian is Equation (29) with real ϕ_A and ϕ_B. Therefore, the eigenvalue equation is as in (30), with $a = E_0 = 48t\phi_A\phi_B + 12t\phi_B^2 - 24V\rho_A\rho_B - 6V\rho_B^2$, $b = E_0 + 12(2V\rho_A + V\rho_B - \mu)$, $c = E_0 + 8(3V\rho_B - \mu)$, and $d = E_0 + 4(6V\rho_A + 9V\rho_B - 5\mu)$; moreover, $u = -12t(2\phi_A + \phi_B)$ and $v = -24t\phi_B$. The minimum eigenvalue is like in the first line of Equation (31), but its explicit form in terms of MF parameters is now

$$E = E_0 + 2(6V\rho_A + 9V\rho_B - 5\mu)$$
$$-6\sqrt{(2V\rho_A + V\rho_B - \mu)^2 + 4t^2(2\phi_A + \phi_B)^2} - 4\sqrt{(3V\rho_B - \mu)^2 + 36t^2\phi_B^2}. \quad (A10)$$

Using the latter formula, the linear system for the coordinates (x, y, z, w) of an eigenvector $|\psi_E\rangle$ of E are:

$$\begin{cases} -(p_1 + p_2)x + q_1 y + q_2 z = -\left(\sqrt{p_1^2 + q_1^2} + \sqrt{p_2^2 + q_2^2}\right) x \\ q_1 x + (p_1 - p_2)y + q_2 w = -\left(\sqrt{p_1^2 + q_1^2} + \sqrt{p_2^2 + q_2^2}\right) y \\ q_2 x - (p_1 - p_2)z + q_1 w = -\left(\sqrt{p_1^2 + q_1^2} + \sqrt{p_2^2 + q_2^2}\right) z \\ q_2 y + q_1 z + (p_1 + p_2)w = -\left(\sqrt{p_1^2 + q_1^2} + \sqrt{p_2^2 + q_2^2}\right) w \end{cases} \quad (A11)$$

with

$$p_1 = 3(2V\rho_A + V\rho_B - \mu); \quad q_1 = -6t(2\phi_A + \phi_B);$$
$$p_2 = 2(3V\rho_B - \mu); \quad q_2 = -12t\phi_B. \quad (A12)$$

Therefore, $|\psi_E\rangle$ is also an eigenvector of a matrix simpler than the original one, with eigenvalue $-\sqrt{p_1^2 + q_1^2} - \sqrt{p_2^2 + q_2^2}$. When $q_1, q_2 \neq 0$, one such vector is:

$$|\psi_E\rangle = \left(\frac{p_1 + \sqrt{p_1^2 + q_1^2}}{q_1} \frac{p_2 + \sqrt{p_2^2 + q_2^2}}{q_2}, -\frac{p_2 + \sqrt{p_2^2 + q_2^2}}{q_2}, -\frac{p_1 + \sqrt{p_1^2 + q_1^2}}{q_1}, 1\right). \quad (A13)$$

Now we are ready to impose self-consistency, which eventually leads to:

$$\rho_A = \frac{1}{2} - \frac{p_2}{2\sqrt{p_2^2 + q_2^2}}; \quad \rho_B = \frac{1}{2} - \frac{p_1}{2\sqrt{p_1^2 + q_1^2}}; \quad \phi_A = -\frac{q_2}{2\sqrt{p_2^2 + q_2^2}}; \quad \phi_B = -\frac{q_1}{2\sqrt{p_1^2 + q_1^2}}, \quad (A14)$$

or explicitly:

$$1 - 2\rho_A = \frac{3V\rho_B - \mu}{\sqrt{(3V\rho_B - \mu)^2 + 36t^2\phi_B^2}}; \quad 1 - 2\rho_B = \frac{2V\rho_A + V\rho_B - \mu}{\sqrt{(2V\rho_A + V\rho_B - \mu)^2 + 4t^2(2\phi_A + \phi_B)^2}};$$
$$\phi_A = \frac{3t\phi_B}{\sqrt{(3V\rho_B - \mu)^2 + 36t^2\phi_B^2}}; \quad \phi_B = \frac{t(2\phi_A + \phi_B)}{\sqrt{(2V\rho_A + V\rho_B - \mu)^2 + 4t^2(2\phi_A + \phi_B)^2}}. \quad (A15)$$

Probably, the simpler method to solve these four coupled non-linear equations in four unknowns is to minimize a suitable non-negative function constructed in such a way as to vanish when the (A15) are all fulfilled. This is easy to do numerically with a computer. By this method, I have drawn the supersolid boundaries in Figure 5.

References

1. Bloch, I.; Dalibard, J.; Zwerger, W. Many-body physics with ultracold gases. *Rev. Mod. Phys.* **2008**, *80*, 885–964. [CrossRef]
2. Amico, L.; Boshier, M.; Birkl, G.; Minguzzi, A.; Miniatura, C.; Kwek, L.-C.; Aghamalyan, D.; Ahufinger, V.; Andrei, N.; Arnold, A.S.; et al. Roadmap on Atomtronics. *arXiv* **2020**, arXiv:2008.04439.
3. Fisher, M.P.A.; Weichman, P.B.; Grinstein, G.; Fisher, D.S. Boson localization and the superfluid-insulator transition. *Phys. Rev. B* **1989**, *40*, 546–570. [CrossRef]
4. Jaksch, D.; Bruder, C.; Cirac, J.I.; Gardiner, C.W.; Zoller, P. Cold Bosonic Atoms in Optical Lattices. *Phys. Rev. Lett.* **1998**, *81*, 3108–3111. [CrossRef]
5. Greiner, M.; Mandel, O.; Esslinger, T.; Hänsch, T.W.; Bloch, I. Quantum phase transition from a superfluid to a Mott insulator in a gas of ultracold atoms. *Nature* **2002**, *415*, 39–44. [CrossRef]
6. van Otterlo, A.; Wagenblast, K.-H.; Baltin, R.; Fazio, R.; Schön, G. Quantum phase transitions of interacting bosons and the supersolid phase. *Phys. Rev. B* **1995**, *52*, 16176–16186. [CrossRef]
7. Griesmaier, A.; Werner, J.; Hensler, S.; Stuhler, J.; Pfau, T. Bose-Einstein Condensation of Chromium. *Phys. Rev. Lett.* **2005**, *94*, 160401. [CrossRef] [PubMed]
8. Lu, M.; Burdick, N.Q.; Youn, S.H.; Lev, B.L. Strongly Dipolar Bose-Einstein Condensate of Dysprosium. *Phys. Rev. Lett.* **2011**, *107*, 190401. [CrossRef] [PubMed]
9. Rokhsar, D.S.; Kotliar, B.G. Gutzwiller projection for bosons. *Phys. Rev. B* **1991**, *44*, 10328–10332. [CrossRef] [PubMed]
10. Krauth, W.; Caffarel, M.; Bouchaud, J.-P. Gutzwiller wave function for a model of strongly interacting bosons. *Phys. Rev. B* **1992**, *45*, 3137–3140. [CrossRef] [PubMed]
11. Sheshadri, K.; Krishnamurthy, H.R.; Pandit, R.; Ramakrishnan, T.V. Superfluid and Insulating Phases in an Interacting-Boson Model: Mean-Field Theory and the RPA. *Europhys. Lett.* **1993**, *22*, 257–263. [CrossRef]
12. Freericks, J.K.; Monien, H. Phase diagram of the Bose-Hubbard Model. *Europhys. Lett.* **1994**, *26*, 545–550. [CrossRef]
13. van Oosten, D.; van der Straten, P.; Stoof, H.T.C. Quantum phases in an optical lattice. *Phys. Rev. A* **2001**, *63*, 053601. [CrossRef]
14. Schroll, C.; Marquardt, F.; Bruder, C. Perturbative corrections to the Gutzwiller mean-field solution of the Mott-Hubbard model. *Phys. Rev. A* **2004**, *70*, 053609. [CrossRef]
15. dos Santos, F.E.A.; Pelster, A. Quantum phase diagram of bosons in optical lattices. *Phys. Rev. A* **2009**, *79*, 013614. [CrossRef]
16. Kübler, M.; Sant'Ana, F.T.; dos Santos, F.E.A.; Pelster, A. Improving mean-field theory for bosons in optical lattices via degenerate perturbation theory. *Phys. Rev. A* **2019**, *99*, 063603. [CrossRef]
17. Batrouni, G.G.; Rousseau, V.; Scalettar, R.T.; Rigol, M.; Muramatsu, A.; Denteneer, P.J.H.; Troyer, M. Mott Domains of Bosons Confined on Optical Lattices. *Phys. Rev. Lett.* **2002**, *89*, 117203. [CrossRef]
18. Capogrosso-Sansone, B.; Prokof'ev, N.V.; Svistunov, B.V. Phase diagram and thermodynamics of the three-dimensional Bose-Hubbard model. *Phys. Rev. B* **2007**, *75*, 134302. [CrossRef]
19. Góral, K.; Santos, L.; Lewenstein M. Quantum Phases of Dipolar Bosons in Optical Lattices. *Phys. Rev. Lett.* **2002**, *88*, 170406. [CrossRef]
20. Kovrizhin, D. L.; Pai, G.V.; Sinha, S. Density wave and supersolid phases of correlated bosons in an optical lattice. *Europhys. Lett.* **2005**, *72*, 162–168. [CrossRef]
21. Sengupta, P.; Pryadko, L.P.; Alet, F.; Troyer, M.; Schmid, G. Supersolids versus Phase Separation in Two-Dimensional Lattice Bosons. *Phys. Rev. Lett.* **2005**, *94*, 207202. [CrossRef] [PubMed]
22. Yamamoto, K.; Todo, S.; Miyashita, S. Successive phase transitions at finite temperatures toward the supersolid state in a three-dimensional extended Bose-Hubbard model. *Phys. Rev. B* **2009**, *79*, 094503. [CrossRef]
23. Pollet, L.; Picon, J.D.; Büchler, H.P.; Troyer, M. Supersolid Phase with Cold Polar Molecules on a Triangular Lattice. *Phys. Rev. Lett.* **2010**, *104*, 125302. [CrossRef] [PubMed]
24. Ng, K.-K. Thermal phase transitions of supersolids in the extended Bose-Hubbard model. *Phys. Rev. B* **2010**, *82*, 184505. [CrossRef]
25. Iskin, M. Route to supersolidity for the extended Bose-Hubbard model. *Phys. Rev. A* **2011**, *83*, 051606(R). [CrossRef]

26. Kimura, T. Gutzwiller study of phase diagrams of extended Bose-Hubbard models. *J. Phys. Conf. Ser.* **2012**, *400*, 012032. [CrossRef]
27. Ohgoe, T.; Suzuki, T.; Kawashima, N. Commensurate Supersolid of Three-Dimensional Lattice Bosons. *Phys. Rev. Lett.* **2012**, *108*, 185302. [CrossRef]
28. Wessel, S.; Troyer, M. Supersolid Hard-Core Bosons on the Triangular Lattice. *Phys. Rev. Lett.* **2005**, *95*, 127205. [CrossRef]
29. Kurdestany, J.M.; Pai, R.V.; Pandit, R. The Inhomogeneous Extended Bose-Hubbard Model: A Mean-Field Theory. *Ann. Phys.* **2012**, *524*, 234–244. [CrossRef]
30. Zhang, X.-F.; Dillenschneider, R.; Yu, Y.; Eggert, S. Supersolid phase transitions for hard-core bosons on a triangular lattice. *Phys. Rev. B* **2011**, *84*, 174515. [CrossRef]
31. Yamamoto, D.; Masaki, A.; Danshita, I. Quantum phases of hardcore bosons with long-range interactions on a square lattice. *Phys. Rev. B* **2012**, *86*, 054516. [CrossRef]
32. Gheeraert, N.; Chester, S.; May, M.; Eggert, S.; Pelster A. Mean-Field Theory for Extended Bose-Hubbard Model with Hard-Core Bosons. In *Selforganization in Complex Systems: The Past, Present, and Future of Synergetics*; Pelster, A., Wunner, G., Eds.; Springer: Zurich, Switzerland, 2016; pp. 289–296.
33. Post, A.J.; Glandt, E.D. Statistical thermodynamics of particles adsorbed onto a spherical surface. I. Canonical ensemble. *J. Chem. Phys.* **1986**, *85*, 7349–7358. [CrossRef]
34. Prestipino Giarritta, S.; Ferrario, M.; Giaquinta, P.V. Statistical geometry of hard particles on a sphere. *Physica A* **1992**, *187*, 456–474. [CrossRef]
35. Prestipino Giarritta, S.; Ferrario, M.; Giaquinta, P.V. Statistical geometry of hard particles on a sphere: Analysis of defects at high density. *Physica A* **1993**, *201*, 649–665. [CrossRef]
36. Prestipino, S.; Speranza, C.; Giaquinta, P.V. Density anomaly in a fluid of softly repulsive particles embedded in a spherical surface. *Soft Matter* **2012**, *8*, 11708–11713. [CrossRef]
37. Vest, J.-P.; Tarjus, G.; Viot, P. Glassy dynamics of dense particle assemblies on a spherical substrate. *J. Chem. Phys.* **2018**, *148*, 164501. [CrossRef]
38. Guerra, R.E.; Kelleher, C.P.; Hollingsworth, A.D.; Chaikin, P.M. Freezing on a sphere. *Nature* **2018**, *554*, 346–350. [CrossRef]
39. Franzini, S.; Reatto, L.; Pini, D. Formation of cluster crystals in an ultra-soft potential model on a spherical surface. *Soft Matter* **2018**, *14*, 8724–8739. [CrossRef]
40. Prestipino, S.; Giaquinta, P.V. Ground state of weakly repulsive soft-core bosons on a sphere. *Phys. Rev. A* **2019**, *99*, 063619. [CrossRef]
41. Prestipino, S.; Sergi, A.; Bruno, E.; Giaquinta, P.V. A variational mean-field study of clusterization in a zero-temperature system of soft-core bosons. *EPJ Web Conf.* **2020**, *230*, 00008. [CrossRef]
42. Zobay, O.; Garraway, B.M. Atom trapping and two-dimensional Bose-Einstein condensates in field-induced adiabatic potentials. *Phys. Rev. A* **2004**, *69*, 023605. [CrossRef]
43. Garraway, B.M.; Perrin, H. Recent developments in trapping and manipulation of atoms with adiabatic potentials. *J. Phys. B Mol. Opt. Phys.* **2016**, *49*, 172001. [CrossRef]
44. Elliott, E.R.; Krutzik, M.C.; Williams, J.R.; Thompson, R.J.; Aveline, D.C. NASA's Cold Atom Lab (CAL): System development and ground test status. *NPJ Microgravity* **2018**, *4*, 16. [CrossRef]
45. Lundblad, N.; Carollo, R.A.; Lannert, C.; Gold, M.J.; Jiang, X.; Paseltiner, D.; Sergay, N.; Aveline, D.C. Shell potentials for microgravity Bose-Einstein condensates. *NPJ Microgravity* **2019**, *5*, 30. [CrossRef]
46. Prestipino, S.; Saija, F. Phase diagram of Gaussian-core nematics. *J. Chem. Phys.* **2007**, *126*, 194902. [CrossRef]
47. Errington, J.R.; Debenedetti, P.G. Relationship between structural order and the anomalies of liquid water. *Nature* **2001**, *409*, 318–321. [CrossRef]
48. Likos, C.N.; Lang, A.; Watzlawek, M.; Löwen, H. Criterion for determining clustering versus reentrant melting behavior for bounded interaction potentials. *Phys. Rev. E* **2001**, *63*, 031206. [CrossRef]
49. Mladek, B.M.; Charbonneau, P.; Likos, C.N.; Frenkel, D.; Kahl, G. Multiple occupancy crystals formed by purely repulsive soft particles. *J. Phys. Condens. Matter* **2008**, *20*, 494245. [CrossRef]
50. Prestipino, S.; Saija, F. Hexatic phase and cluster crystals of two-dimensional GEM4 spheres. *J. Chem. Phys.* **2014**, *141*, 184502. [CrossRef]
51. Kunimi, M.; Kato, Y. Mean-field and stability analyses of two-dimensional flowing soft-core bosons modeling a supersolid. *Phys. Rev. B* **2012**, *86*, 060510(R). [CrossRef]

52. Macrì, T.; Maucher, F.; Cinti, F.; Pohl, T. Elementary excitations of ultracold soft-core bosons across the superfluid-supersolid phase transition. *Phys. Rev. A* **2013**, *87*, 061602(R). [CrossRef]
53. Prestipino, S.; Sergi, A.; Bruno, E. Freezing of soft-core bosons at zero temperature: A variational theory. *Phys. Rev. B* **2018**, *98*, 104104. [CrossRef]

Publisher's Note: MDPI stays neutral with regard to jurisdictional claims in published maps and institutional affiliations.

© 2020 by the author. Licensee MDPI, Basel, Switzerland. This article is an open access article distributed under the terms and conditions of the Creative Commons Attribution (CC BY) license (http://creativecommons.org/licenses/by/4.0/).

Article

Real Space Triplets in Quantum Condensed Matter: Numerical Experiments Using Path Integrals, Closures, and Hard Spheres

Luis M. Sesé

Departamento de Ciencias y Técnicas Fisicoquímicas, Facultad de Ciencias, Universidad Nacional de Educación a Distancia (UNED), Avda. Esparta s/n, 28232 Las Rozas, Madrid, Spain; msese@ccia.uned.es

Received: 28 October 2020; Accepted: 21 November 2020; Published: 25 November 2020

Abstract: Path integral Monte Carlo and closure computations are utilized to study real space triplet correlations in the quantum hard-sphere system. The conditions cover from the normal fluid phase to the solid phases face-centered cubic (FCC) and cI16 (de Broglie wavelengths $0.2 \leq \lambda_B^* < 2$, densities $0.1 \leq \rho_N^* \leq 0.925$). The focus is on the equilateral and isosceles features of the path-integral centroid and instantaneous structures. Complementary calculations of the associated pair structures are also carried out to strengthen structural identifications and facilitate closure evaluations. The three closures employed are Kirkwood superposition, Jackson–Feenberg convolution, and their average (AV3). A large quantity of new data are reported, and conclusions are drawn regarding (i) the remarkable performance of AV3 for the centroid and instantaneous correlations, (ii) the correspondences between the fluid and FCC salient features on the coexistence line, and (iii) the most conspicuous differences between FCC and cI16 at the pair and the triplet levels at moderately high densities ($\rho_N^* = 0.9, 0.925$). This research is expected to provide low-temperature insights useful for the future related studies of properties of real systems (e.g., helium, alkali metals, and general colloidal systems).

Keywords: quantum triplets; path integral Monte Carlo; closures; quantum hard spheres; fluid–solid transition; FCC solid; cI16 solid

1. Introduction

The study of equilibrium triplet structures in 3D N-particle systems with quantum behavior remains a pending task in condensed matter research at low temperatures. Apart from a number of early developments focused mainly on the proposal and indirect testing of the so-called *closures* [1–11] or on the use of alternative order parameters [12], just a few computational works based on Feynman's path integrals (PI) [13] can be found in the recent literature on this field [14–18]. Not much is known about the behavior of quantum triplets, hence the interest in undertaking this task. This is not only a logical step further in current statistical mechanics, allowing one to formulate thermodynamic properties beyond the pairwise approach [19], but also it is central to outstanding condensed matter properties. Among the latter, one can mention the following: phonon–phonon interactions in helium-II [4], the N-particle interpretation of fluid entropies [20–22], multiple scattering [23], theories of phase transitions [24,25], and glassy dynamics [26–28]. Although the whole PI quantum triplet task is computationally daunting at the present time, one can always seek to identify the main triplet features that may serve as a guide for the necessary future work on this topic.

The triplet topic encompasses both real-space $\{r\}$–triplets and reciprocal (Fourier)-space $\{k\}$–triplets. Nevertheless, there is no direct experimental determination of triplet functions [29,30]. Thus, one must resort to theoretical computations for extracting this sort of information, which makes these computations the only "experimental" method of solving the triplet problem. In the quantum case, a rigorous framework for triplets is given by PI, albeit the computations are truly exacting [14–18].

In this regard, and leaving aside exchange interactions for simplicity, one notes that just the quantum thermal delocalization of the particles is sufficient to bring about a much higher complexity in the quantum study than that present in the classical domain. This helps to understand the key role in this topic that was played by closures, which represent fluid triplet correlation functions $g_3(r, s, u)$ utilizing pair information $g_2(r)$, thus involving affordable computations. Noticeable among the closures for the fluid triplets $g_3(r, s, u)$ are the early Kirkwood superposition KS3 [1,3], and the key Jackson–Feenberg convolution JF3 [4,24], although other forms with even a wider scope are available (e.g., triplet direct correlation functions $c_3(r, s, u)$) [5,24,29]. Despite the fact that closures are approximations to the actual fluid triplet functions, they are still highly valuable in that they may provide insightful physical pictures of the underlying structure of the triplet correlations. Therefore, even nowadays, closures should not be disregarded without giving them the opportunity to prove their worth as interpretative tools in the quantum domain [17,18].

The PI formalism is perfectly suited for performing path integral Monte Carlo (PIMC) and molecular dynamics (PIMD) computer simulations of quantum N-particle systems at nonzero temperatures [31–55]. With due attention to the special characteristics of quantum averages [33,38,44–47], the latter PI simulations can follow classical-like procedures [56–60]. To illustrate this situation, it is sufficient to recall the most basic PI description in the canonical ensemble (N, V, T) of an actual quantum monatomic system in which exchange interactions can be neglected. Such an actual system is represented by a PI model composed of N necklaces with P beads apiece, the whole set of $N \times P$ beads obeying a classical-like partition function [31–33]. It is important to note that P is an integer number, $P > 1$, which is to be optimized to obtain statistical convergence for the properties. (The actual quantum system is retrieved in the Trotter's limit $P \to \infty$ [61], while the classical limit is $P = 1$). Special techniques are available to improve the P description and reduce computations (e.g., pair actions and higher-order propagators [33–35,37,38] combined with algorithms for moving the beads) [33,41,47,62,63]. In this connection, depending on the technique selected, there may or may not exist equivalence between all the beads in the model sample, which is a fact that turns out to be crucial for the study of structures [33,35,38,47]. Thus, one speaks in this context of the structurally significant (equivalent among themselves) beads; their number, say X, takes the convenient values P or $P/2$.

The PI applicability covers from quantum diffraction effects (PIMC and PIMD for atomic and molecular fluids and solids [64–69]) to bosonic quantum exchange (PIMC) [33,36,41,43]. PIMC and PIMD are said to be "exact" in that they produce results with controllable statistical errors. These results have been proven to be in excellent agreement with experimental data [33,41,42,48]. In addition to this, PIMC approaches to fermionic exchange can also be devised [70], although the "sign problem" precludes the related PIMC applications from being definitive. These facts, together with the PI flexibility, make PIMC and PIMD most powerful tools in quantum condensed matter research.

By focusing attention on quantum monatomic systems at equilibrium, with diffraction effects dominating their behavior, it is worthwhile to specify the three general categories of physically significant structures [31,33,38,39,47] that are revealed by PI: (a) instantaneous; (b) total thermalized-continuous linear response; and (c) centroids (centroid = proper center-of-mass of a PI necklace) [38,47]. For each of these categories, three points must be highlighted [47]. First, a given category is associated with the linear response from the system to a distinct weak external field: the instantaneous case is associated with a δ–localizing field (e.g., as in elastic neutron scattering), this usually being the category linked to "the structure" of a quantum system; the total thermalized-continuous linear response with, for example, a continuous field; and the centroids with specifically a constant-strength field. Second, a given category can be formulated in a two-fold way (real space and Fourier space), which extends over the corresponding n-body functions (g_n correlation functions and $S^{(n)}$ static structure factors). Third, a given category is defined by special forms of the averages over the NX bead positions. These averages scale with X in different forms: the instantaneous with X, the total thermalized-continuous linear response with X^n, and the centroids with $X^0 = 1$. In stark contrast, the classical system only shows one category of a physically significant structure [56].

Therefore, it is easy to understand the much higher complexity and computational cost of the quantum problem when treated in depth.

The aim of the present article is to expand the study of the PI triplet features in many-body quantum systems [14–18]. The system selected is that composed of quantum hard spheres (QHS system hereafter). Hard spheres are known to be a very useful reference. They have been used to model classical and quantum systems, ranging from helium [33,36,66,67,71–73] to complex colloids [74,75], and under very different conditions (i.e., fluid, boson superfluid, superlattices, solids, and suspensions). This work concentrates on the real space instantaneous and centroid triplets, leaving aside the total thermalized-continuous linear response case to keep the related computations affordable [17]. As stressed earlier, in understanding triplets through closures, a thorough consideration of the pair structures is needed. This benefits the triplet-closure computations and the analysis of the correspondences between the salient pair and triplet features (in different phases or within the same phase). Therefore, the necessary attention is also paid to the pair prerequisite.

The exact computational method chosen is PIMC, which avoids the PIMD difficulties linked to the discontinuity of the hard-sphere potential, and it is complemented with the closures: KS3, JF3, and their average AV3 = (KS3 + JF3)/2. A wide range of QHS fluid and solid conditions, within the purely diffraction regime, is studied: reduced de Broglie wavelengths $0.2 \leq \lambda_B^* < 2$ and reduced number densities $0.1 \leq \rho_N^* \leq 0.925$, where $\lambda_B^* = h/(2\pi M k_B T \sigma^2)^{1/2}$, $\rho_N^* = N\sigma^3/V$, σ and M being the hard-sphere diameter and mass, respectively. The specific $\{r\}$-space targets pursued are the following:

(i) Analyzing in detail the potential usefulness of AV3 for quantum fluid triplet studies. This is prompted by the encouraging results obtained in [17] for liquid para-hydrogen and liquid neon.

(ii) Gaining triplet structural insights from the comparison, in the short–medium range of distances, between the coexisting fluid and FCC (face-centered cubic) solid [66,67].

(iii) Comparing the salient triplet features of the cubic solids FCC and cI16 at moderately high densities, ($\rho_N^* = 0.925$, $\lambda_B^* = 0.2$) and ($\rho_N^* = 0.9$, $\lambda_B^* = 0.8$). The so-called cI16 lattice is a distorted superstructure of the body-centered cubic (BCC) lattice [76,77]. (Hard-sphere BCC lattices are known to be mechanically unstable in both classical and quantum applications [78,79]). One also notes that cI16 phases have been reported for Li and Na at very high pressures [76,77], hence the interest of this point.

(iv) In connection with (iii), there is the question of establishing a clear identification of the QHS bcc-qIII phases, observed in [67,78], as genuine cI16 phases. (The bcc-qIII phases arise from the PIMC evolution of initially perfect BCC lattices). The cI16 lattice has been identified recently for classical hard spheres in the insightful simulation work reported in [80], and the patterns of the related results suggest that bcc-qIII is in fact cI16. Proof of this is given in this article, which adds more value to the QHS system for modeling quantum solid–solid changes of phase [67] and enhances the meaning of the related triplet calculations.

It is hoped that the reported "experimental" results will form a useful basis for comparison when extensive studies of triplets in real quantum systems are undertaken. The outline of this article is as follows. Section 2 contains the theoretical background. Section 3 describes the relevant computational details. Section 4 gives the results and their discussion, and Section 5 collates the main conclusions of this work.

2. Theory

2.1. Path Integral Monte Carlo (PIMC)

The canonical partition function $Z(N, V, T)$ of a quantum monatomic system (number density $\rho_N = N/V$), under conditions where exchange interactions can be neglected, can be approximated by the accurate PI formula (Tr = trace) [31–33]

$$Z = Tr\{\exp(-\beta H_N)\} \cong \frac{1}{N!}\left(\frac{MP}{2\pi\beta\hbar^2}\right)^{3NP/2} \int \prod_{j=1}^{N}\prod_{t=1}^{P} dr_j^t \times \exp[-\beta W_{NP}], \quad (1)$$

where H_N is the Hamiltonian, assumed normally to be composed of one- and two-particle terms, M is the particle mass, $\beta = 1/k_B T$, P is the discretization in beads of the necklace representing and actual particle j, r_j^t denotes the real space coordinates of bead t ($t = 1, 2, \ldots, P$) belonging to necklace j, and W_{NP} is the "effective potential" ruling the whole set of $N \times P$ beads (hereafter all of them equivalent: $X = P$). In what follows, the optimal P will be assumed. In addition, it is worthwhile to note that (a) consecutive beads, t and $t + 1$, in a necklace are separated by $\beta\hbar/P$ in Euclidean imaginary time $\beta\hbar$; (b) then, a given bead labeled t is associated with the imaginary time $\beta\hbar t/P$; and (c) the cyclic property $t + 1 = P + 1 \equiv 1$ is satisfied.

In the case of the QHS system, an appropriate choice for W_{NP} is based on Cao–Berne's CBHSP pair action [35], and it can be written as [14,78]

$$W_{NP}^{CBHSP} = W_1^F + W_2^{CB} + W_2^{HS}, \quad (2)$$

$$W_1^F = \frac{MP}{2\beta^2\hbar^2}\sum_{j=1}^{N}\sum_{t=1}^{P*}\left(r_j^t - r_j^{t+1}\right)^2, \quad (3a)$$

$$W_2^{CB} = -\beta^{-1}\ln\prod_{j<m}\prod_{t=1}^{P*}\left\{1 - \frac{\sigma\left(r_{jm}^t + r_{jm}^{t+1} - \sigma\right)}{r_{jm}^t r_{jm}^{t+1}} \times \exp\left[-\frac{MP\left(r_{jm}^t - \sigma\right)\left(r_{jm}^{t+1} - \sigma\right)}{2\beta\hbar^2}\left(1 + \cos(r_{jm}^t, r_{jm}^{t+1})\right)\right]\right\}, \quad (3b)$$

$$W_2^{HS} = P^{-1}\sum_{j<m}^{P}\sum_{t=1}^{}\omega_{HS}(r_{jm}^t) = \begin{cases} \infty \text{ if } & r_{jm}^t = \left|r_j^t - r_m^t\right| < \sigma \\ 0 \text{ if } & r_{jm}^t > \sigma \end{cases}. \quad (3c)$$

In Equation (3a), one recognizes the superposition of the free-particle behaviors [13]. Equation (3b) shows Cao–Berne's correction, where the adjacent-bead effects are to be noted. Equation (3c) is the expression of the singular hard-sphere potential extended over all the pairs of necklaces, which interact in an equal "t–time" bead-to-bead fashion (ET). The symbols $P*$ in the sum and product above mark the t–cyclic property already mentioned. For the specific thermodynamic property formulas that can be derived from CBHSP, the reader is referred to [67,78]. At this point, it is important to give the definition of the CBHSP centroid of a given necklace j

$$R_{CM,j} = P^{-1}\sum_{t=1}^{P} r_j^t. \quad (4)$$

This quantity plays an important role in PI theoretical developments [13,47], in particular in (a) the appealing centroid approaches to quantum dynamics [81–85]; (b) the exact formulation of the equation of state of quantum fluids [39,86]; and (c) the characterization of quantum solid phases [67].

A key feature of the QHS system is that its state points can be uniquely characterized by giving two parameters, namely the reduced number density ρ_N^* and the reduced de Broglie thermal wavelength λ_B^*. This fact was early realized at the level of semiclassical treatments based on $\hbar$–expansions (see, for instance, [87–90]). Within PI, the same fact is just a property contained in the mathematical structure of the QHS partition function, regardless of the particular proper form that W_{NP} may take (see [47] for a discussion of QHS propagators). Accordingly, the QHS system properties can be expressed in reduced units, thereby being independent of the actual parameters M, σ, T, and ρ_N employed [66,67]. Therefore, for example, internal energies E can be given as $E^* = E/RT$, and pressures p can be given as $p^* = pM\sigma^5/\hbar^2$. For the pressure, an indication to guide the interested reader will suffice: when using different sets of parameters to define the state points 1 and 2, if $(\rho_N^*, \lambda_B^*)_1 = (\rho_N^*, \lambda_B^*)_2$, then necessarily $(PV/RT)_1 = (PV/RT)_2$, and also $p_1^* = p_2^*$.

The same general type of argument applies to the real space structures $g_n(q_1, q_2, \ldots, q_n)$, for which, when reporting QHS system results, it is most useful to do it using interparticle distances in reduced form: $r_{12}^* = |q_1 - q_2|/\sigma$. In order not to burden the notation, the formulation of the structural concepts below will utilize the distances and related quantities in their non-reduced forms, as in Equations (1) to (4).

Another technical point seems worth placing here. It is related to the three- (and higher-order) particle contributions to the quantum Hamiltonian H_N of the system, which may yield more complete and elaborate forms for the propagators and W_{NP}. While this is a question that can be tackled in various ways when continuous interparticle potentials are involved [91–93], to this author's knowledge, no QHS system PI actions beyond the pair level are available, and such an extension remains intractable for now. Nevertheless, because of the strong similarity between helium atoms and quantum hard spheres, the related effects on the QHS system can be expected to become significant at very high solid-phase densities (and sufficiently low temperatures) [33]. Furthermore, owing to the QHS infinite repulsion at the hard core, Equation (3c), the wave functions of the QHS system must vanish for interparticle distances $r \leq \sigma$ (i.e., there can be no tunneling); hence, quantum hard spheres repel one another before "classical contact" can occur [89,90]. (Within PI, this means that the related forbidden region brought about by Equation (3c) arises only for the "equal-time" bead g_n correlations). Therefore, given the lack of any attraction, the "preemptive" QHS repulsions can be expected to cause a strong impediment to the coming together of triplets of quantum hard spheres. Using the quantum diffraction parameter $\gamma = \rho_N^* \lambda_B^{*3}$, the latter triplet effects should not play any significant role unless γ becomes truly high. The largest value of γ in this work is $\simeq 2.8$, which is compatible with the QHS pair modeling of normal fluid and solid helium-4 [66]. Therefore, the pair-level CBHSP approach can be deemed adequate to compute structures under the fluid and solid conditions investigated in this work.

2.2. PI Triplet Structures

Within PI-CBHSP, the centroid (CM3) and the instantaneous (ET3) three-point number densities can be cast as the ensemble averages [17]

$$\rho_{CM3}^{(3)}(q_1, q_2, q_3) = \langle \sum_{j \neq l \neq m \neq j} \delta(R_{CM,j} - q_1) \delta(R_{CM,l} - q_2) \delta(R_{CM,m} - q_3) \rangle, \tag{5}$$

$$\rho_{ET3}^{(3)}(q_1, q_2, q_3) = P^{-1} \langle \sum_{t=1}^{P} \sum_{j \neq l \neq m \neq j} \delta(r_j^t - q_1) \delta(r_l^t - q_2) \delta(r_m^t - q_3) \rangle, \tag{6}$$

where one notices that (i) the multi-index summations run over the whole set of permutations of N particles taken three at a time; (ii) the instantaneous case contains a further P average involving "equal-time" beads in different necklaces; and (iii) these definitions are completely general, since they depend on the position vectors of the representative set of three particles and can be applied to the statistical description of monatomic systems, which are either fluid or solid. Due to the high computational cost, no attempt is made in this work at calculating total thermalized-continuous linear response triplets [14,17].

For homogeneous and isotropic fluids, one finds simpler formulas [17]

$$\rho_{CM3}^{(3)}(q_1, q_2, q_3) = \rho_N^3 g_{CM3}(q_1 - q_3, q_2 - q_3) = \rho_N^3 g_{CM3}(r_{12}, s_{13}, u_{23}), \tag{7}$$

$$\rho_{ET3}^{(3)}(q_1, q_2, q_3) = \rho_N^3 g_{ET3}(q_1 - q_3, q_2 - q_3) = \rho_N^3 g_{ET3}(r_{12}, s_{13}, u_{23}), \tag{8}$$

where the triplet correlation functions g_{CM3} and g_{ET3} depend only on the three generic interparticle distances: $r_{12} = |q_1 - q_2|$, $s_{13} = |q_1 - q_3|$, and $u_{23} = |q_2 - q_3|$. This exact reduction from nine to three independent variables makes the intricate triplet problem more accessible for the study of monatomic fluid state points [14–17].

Rigorously speaking, the related exact framework for a monatomic solid is contained in Equations (5) and (6). Nevertheless, affordable approximations to this even more costly problem can be obtained by applying Equations (7) and (8). Actually, such an approach is consistent with the same idea, already exploited successfully, at the pair level in the study of regular solid phases, since the $g_{CM2}(r)$ and $g_{ET2}(r)$ pair radial functions retain many significant traits of the underlying solid structure [66,67,80]. Furthermore, as a first step, the use of Equations (7) and (8) facilitates the comparison of the global salient triplet features appearing in different solid phases.

The functions defined in Equations (7) and (8) must satisfy several properties [4,6,7,57,58]. The most relevant to this work are:

(1) Symmetry

$$g_3(q_1, q_2, q_3) = g_3(q_2, q_3, q_1) = \ldots; \quad \text{ET3 and CM3.} \quad (9a)$$

(2) QHS instantaneous behavior at close distances

$$\lim_{|q_j - q_m| \to \sigma^+} g_{ET3}(r, s, u) = 0. \quad (9b)$$

(3) Asymptotic behavior in fluids

$$\lim_{r \to \infty} g_3(r, r, r) = 1; \quad \text{ET3 and CM3,} \quad (9c)$$

$$\lim_{s \to \infty} g_3(r, s, s) = g_2(r); \quad \text{ET3/ET2 and CM3/CM2.} \quad (9d)$$

Equation (9a) follows from Equations (7) and (8). Equation (9b) for the instantaneous case arises from the singular character of the hard-sphere potential Equation (3c). For centroids, a behavior similar to Equation (9b) is expected to occur, albeit the limiting distance may be different from σ. Finally, Equations (9c) and (9d) follow from the weakening of particle correlations in fluids when considering increasing distances, and both are very useful to check the inner consistency of the related calculations.

2.3. Additional Pair Structural Quantities

To supplement the PIMC triplet calculations in the canonical ensemble, the following quantities can also be computed:

(a) The pair radial functions for the centroid (CM2) and the instantaneous (ET2) correlations, in both the fluid and the solid phases [47]. Their PI ensemble averages can be cast as

$$\rho_N^2 g_{CM2}(r_{12}) = \langle \sum_{j \neq m} \delta(R_{CM,j} - q_1) \delta(R_{CM,m} - q_2) \rangle, \quad (10)$$

$$\rho_N^2 g_{ET2}(r_{12}) = P^{-1} \langle \sum_{t=1}^{P} \sum_{j \neq m} \delta(r_j^t - q_1) \delta(r_m^t - q_2) \rangle, \quad (11)$$

where $r_{12} = |q_1 - q_2|$.

(b) The pair static structure factors $S_{CM}^{(2)}(k)$ and $S_{ET}^{(2)}(k)$ associated with the foregoing pair radial structures in the fluid phase [47]

$$S_{CM}^{(2)}(k) = 1 + \rho_N \int dr_{12} \exp(i k \cdot r_{12}) h_{CM2}(r_{12}) = \left(1 - \rho_N c_{CM}^{(2)}(k)\right)^{-1}, \quad (12)$$

$$S_{ET}^{(2)}(k) = 1 + \rho_N \int dr_{12} \exp(i\mathbf{k}\cdot\mathbf{r}_{12}) h_{ET2}(r_{12}) \cong \left(1 - \rho_N c_{ET}^{(2)}(k)\right)^{-1}, \tag{13}$$

where $h_2 = g_2 - 1$ stands for the corresponding pair *total* correlation function, and $c^{(2)}(k)$ stands for the corresponding pair *direct* correlation function in Fourier space ($k = |\mathbf{k}|$). These structure factors can be fixed with great accuracy, at a very low cost and for every $k \geq 0$ wave number [48], via the Ornstein–Zernike framework [94–96] developed by this author [47,86,97,98]. Apart from their intrinsic usefulness, they are decisive in achieving a number of significant improvements in the study of fluids with quantum behavior [39,47–49]. In particular, $S_{CM}^{(2)}(k)$ and $S_{ET}^{(2)}(k)$ can be utilized for (i) extending the ranges of the simulated $g_{CM2}(r_{12})$ and $g_{ET2}(r_{12})$ [47], which serves to perform triplet closure computations; and (ii) gaining insight into their associated triplet structure factors $S_{CM}^{(3)}(\mathbf{k}_1, \mathbf{k}_2)$ and $S_{ET}^{(3)}(\mathbf{k}_1, \mathbf{k}_2)$ [17,18].

(c) In simulation work using cubic boxes, the PI sample size is composed of N_S necklaces, each with P beads, enclosed in a volume $V_S = L^3$. To characterize solid phases, one can employ the normalized-to-unity solid-phase configurational structure factors at the centroid and instantaneous pair levels [67,99,100]. They can be written as

$$S_{CM2}^{(C)}(\mathbf{k}) = N_S^{-2} \left|\sum_{j=1}^{N_S} \exp(i\mathbf{k}\cdot\mathbf{R}_{CM,j})\right|^2, \tag{14}$$

$$S_{ET2}^{(C)}(\mathbf{k}) = (N_S^2 P)^{-1} \sum_{t=1}^{P} \left|\sum_{j=1}^{N_S} \exp(i\mathbf{k}\cdot\mathbf{r}_j^t)\right|^2, \tag{15}$$

and are taken at their maximal values arising from the simulation runs [67,78]. In these simulation conditions, the wave vectors $\mathbf{k}$ to be analyzed must be commensurate with the box, which means $\mathbf{k} = 2\pi L^{-1}(k_x, k_y, k_z)$, where the components (k_x, k_y, k_z) take integer values [56]. In connection with this, notice that cubic-based *perfect* lattices can be associated with sets of three commensurate wave vectors, $\{\mathbf{k}_w\}_n = \{\mathbf{k}_1, \mathbf{k}_2, \mathbf{k}_3\}_n$, which are maximal in that:

(i) For the perfect FCC and BCC lattices, one can single out sets $\{\mathbf{k}_w\}_n$ such that they reach the maximum value, $S_2^{(C)}(\mathbf{k}_w) = S_2^{(C)}(\mathbf{k}_{max}) = 1, (w = 1, 2, 3)$. For a perfect cI16 lattice, which is not so highly regular, one obtains $S_2^{(C)}(\mathbf{k}_w) = S_2^{(C)}(\mathbf{k}_{max}) < 1, (w = 1, 2, 3)$, as will be shown later on.

(ii) The following result holds

$$|\mathbf{k}_1 \cdot (\mathbf{k}_2 \times \mathbf{k}_3)| = (2\pi)^3 N_S / V_S. \tag{16}$$

Therefore, comparison of the above standard perfect-lattice results with those arising from the simulated cubic solid phase allows one to identify its type and relative order. Obviously, the values of the simulated configurational structure factors are lower than the perfect reference values; they appear associated with each of the three maximal wave vectors and are close to one another, but, as a rule, they are not equal: one of them can be singled out as the maximum, whereas the other two remain slightly below [67,78]. As a guide for quantum work [99], the following *centroid* estimates are worth quoting: $0.4 \lesssim S_{CM2}^{(C)}(\mathbf{k}_{max})$ for partially crystalline solids, while typically $S_{CM2}^{(C)}(\mathbf{k}_{max}) < 0.2$ for fluid phases. (Amorphous phase maximal values for $S_{CM2}^{(C)}(\mathbf{k}_{max})$ should be between the two foregoing limits). It is important to stress that although somewhat expensive to calculate, the quantities $S_{CM2}^{(C)}(\mathbf{k}_{max})$ and $S_{ET2}^{(C)}(\mathbf{k}_{max})$ are global for the simulation sample. Therefore, in this context, these quantities seem more complete than local-order parameters (e.g., the rotationally invariant Q_l) [67,80,101].

Before going any further, it is convenient to consider the general issue of the simulation sample size N_S for the solid phases, thus allowing one to introduce cI16 basic details. The conditions for FCC and BCC are well-known, and for $N_S > 100$, one finds: (i) $N_S(\text{FCC}) = 4n^3$, with $n = 3, 4, 5, \ldots$; and (ii) $N_S(\text{BCC}) = 2n^3$, with $n = 4, 5, 6, \ldots$. However, the case of cI16 is not so standard. cI16 is a distortion of BCC and is characterized by the so-called *fractional displacement parameter*, which is usually denoted by x [76,77], so as to have the particles occupying the 16c Wyckoff site (x,x,x) of the space group $\bar{I}43d$. This means that its body-centered unit cell does contain 16 particles. Consequently, there are some extra restrictions that may make the $N_S(cI16)$ values different from those of BCC. Thus, again for $N_S > 100$, one finds (iii) $N_S(cI16) = 16n^3$, with $n = 2,3,4,\ldots$. The reader is referred to [76,77,80] for specific details.

2.4. Closures for Fluid Triplets

The two basic closures analyzed in this work are Kirkwood superposition KS3 and Jackson–Feenberg convolution JF3. Both can be applied to the fluid centroid (CM3) and instantaneous (ET3) triplet correlations. Their expressions can be written as follows [1,4]:

$$g_{KS3}(r_{12}, s_{13}, u_{23}) = g_2(r_{12}) g_2(s_{13}) g_2(u_{23}), \tag{17}$$

$$g_{JF3}(r_{12}, s_{13}, u_{23}) = g_{KS3}(r_{12}, s_{13}, u_{23}) - h_2(r_{12}) h_2(s_{13}) h_2(u_{23}) + \rho_N \int dq_4 h_2(v_{14}) h_2(v_{24}) h_2(v_{34}), \tag{18}$$

where $v_{j4} = |q_j - q_4|$, $h_2 = g_2 - 1$, and $g_2 = g_{CM2}$ or g_{ET2}. Although explicitly stated in Equation (18), it is important to remark that JF3 lacks the triplet-product term $h_2(r_{12}) h_2(s_{13}) h_2(u_{23})$, which should appear in an h_2-expansion. This absence has deep consequences as will be shown in this article. An easy and direct way to recover such contribution (half of it), while at the same time keeping the convolution integral (half of it) contained in Equation (18), is via the average closure AV3 that reads as

$$g_{AV3}(r_{12}, s_{13}, u_{23}) = \frac{1}{2}\bigl(g_{KS3}(r_{12}, s_{13}, u_{23}) + g_{JF3}(r_{12}, s_{13}, u_{23})\bigr). \tag{19}$$

As regards the properties of these closures, suffice it to say that (i) KS3, JF3, and AV3, satisfy Equations (9a), (9c), and (9d); and (ii) only KS3, as induced by its construction, satisfies Equation (9b), which is a special case of the general triplet behavior $g_3 \to 0$ when two particles approach closely each other [14–17].

3. Computational Details

The main target of this work is the determination of QHS equilateral and isosceles triplet correlations (centroid and instantaneous), namely the types of functions $g_3(r,r,r)$ (or g_3^{EQ} for brevity when necessary) and $g_3(r,s,s)$. For the sake of interpretation, they are complemented with the additional structural properties discussed in Section 2.3. The state points studied are shown in Table 1. They span a wide range of conditions, from the normal fluid phase to the distinct solid phases FCC and cI16. Special attention is paid to the two sides of the fluid–FCC coexistence line, as determined in [67] $(\lambda_B^* \leq 0.8)$ and [66] $(\lambda_B^* > 0.8)$. Moreover, the study is extended to (i) fluid state points under very strong diffraction effects $(\lambda_B^* \approx 2)$, with a view to establishing a meaningful correlation of triplet structures when going toward the change of phase by increasing ρ_N^* at constant temperature, and (ii) the lattices FCC and cI16 at $(\rho_N^* = 0.925, \lambda_B^* = 0.2)$ and $(\rho_N^* = 0.9, \lambda_B^* = 0.8)$, which are conditions that are significantly far from the very high-density regions.

Table 1. Fluid and solid-state points of the hard-sphere system studied. Reduced densities ρ_N^*, reduced de Broglie wavelengths λ_B^*, path integral Monte Carlo (PIMC) sample size $N_S \times P$.

	I.1. PHASE TRANSITION [1]				
	FLUID PHASE			FCC PHASE	
λ_B^*	ρ_N^*	$N_S \times P$		ρ_N^*	$N_S \times P$
0.2	0.789	864×12		0.863	864×12
0.4	0.672	864×12		0.731	864×12

Table 1. Cont.

		I.1. PHASE TRANSITION [1]		
	FLUID PHASE		**FCC PHASE**	
λ_B^*	ρ_N^*	$N_S \times P$	ρ_N^*	$N_S \times P$
0.6	0.589	864 × 12	0.635	864 × 12
0.8	0.533	864 × 12 864 × 24	0.573	864 × 12 864 × 18
1.2543	0.442	864 × 24	0.465	864 × 24
1.9832	0.348	864 × 30 864 × 40	0.360	864 × 30
		I.2. FLUID PHASE		
1.9832	0.1	864 × 30	———	———
1.9832	0.3	864 × 30	———	———
		I.3 SOLID PHASES		
	cI16 PHASE		**FCC PHASE**	
0.2	0.925	1024 × 12 1024 × 24	0.925	864 × 12 864 × 24
0.8	0.900	1024 × 12 1024 × 24 1024 × 36	0.900	864 × 12 864 × 24 864 × 36

[1] Phase transition de Broglie wavelengths and densities fixed in [66,67].

3.1. PIMC Calculations

The PIMC simulation procedures utilized can be found elsewhere [14–17,67,78], although for completeness, the basic lines follow below.

The necklace normal mode algorithm [62,63] is used to generate the collective P movements of a given necklace, with a Metropolis acceptance ratio of 50%. (As in previous works, the actual hard-sphere parameters are $M = 28.0134$ amu and $\sigma = 3.5$ Å; 1 Å = 10^{-10} m). The necklace sample sizes N_S are 864 for the fluid and the FCC solid phases, and 1024 for the cI16 solid phases. The quantum P convergence for the results is studied as shown below ($12 \leq P \leq 40$). One kpass is defined as a set of $10^3 N_S \times P$ attempted bead moves, and one Mpass is then 10^3 kpasses. After equilibration, most of the simulation runs are arranged into 40 blocks for the g_2 calculations and 30 blocks for the g_3 calculations. The respective block sizes are (i) 92.6 kpasses for the fluid simulations; (ii) 92.6 kpasses for the FCC simulations; and (iii) 78.125 kpasses for the cI16 simulations. Therefore, the run lengths associated with the g_2 and g_3 calculations are in between 2.34 Mpasses and 3.7 Mpasses. (The extra simulations using $P = 36$ and 40 have lengths of about 1 Mpass). Block subaverages for g_2 and g_3 are obtained by gathering statistics every 5000 (g_2)/7000–8000 (g_3) configurations generated. The configurational structure factors given by Equations (14) and (15) are analyzed four times per block, at equally spaced intervals, by recording the ten largest values for the final assessment. To do so, triplets of integers (k_x, k_y, k_z) are monitored in the mesh $25 \leq k_x^2 + k_y^2 + k_z^2 \leq 200$, with the components in $-10 \leq k_\nu \leq 10$ (symmetry properties allow one to reduce the calculations). Given that the information provided by the correlation functions, complemented with that arising from the structure factors, is more than sufficient to characterize the current solid structural results, the Q_l order parameters [101] are not evaluated, thereby alleviating the considerable computational effort involved in this work.

The pair and triplet sructures g_2 and g_3 are fixed in the established ways using histograms. The case of g_2 is straightforward and well-known [56], and the simulations are utilized as the reduced width of the bins $\vec{\Delta}^* = 1/35$ (or $\sigma/35 = 0.1$ Å). However, the case of g_3 includes a good number of subtleties [57,58]. The detailed description of the related procedure can be found in [14]. For the current purposes, suffice it to say that for a triplet of distances (R, S, U), the basic g_3 expression is given by

$$g_3(R, S, U) = \frac{(\Delta n_T)}{N \rho_N^2 (\Delta V)_{RSU}}; \text{ ET3 and CM3} \qquad (20)$$

where (Δn_T) is the number of times mutual distance triplets lie within the ranges $R - \overline{\Delta} < r_{12} \leq R$, $S - \overline{\Delta} < s_{13} \leq S$, and $U - \overline{\Delta} < u_{23} \leq U$, and $(\Delta V)_{RSU}$ stands for the appropriate volume element [58]. Once again, in these calculations, $\overline{\Delta}^* = 1/35$. The histogramming of triplets extends up to distances r_{12}, s_{13}, and u_{23}, which are $< L/4$. Statistical errors (one-standard deviation) for the average structures computed with PIMC are fixed with the corresponding block subaverages. For example, for the first peaks heights of g_2 and g_3, the errors remain below 1% for most of the present calculations. In this connection, Table 2 gives some representative g_3 results (mean first peaks (FP)) in the close vicinities of the absolute maxima of the structure indicated, together with the associated errors. (More on this in the Supplementary Materials). Note that the P convergence is influenced by both λ_B^* and ρ_N^*. For the fluid and FCC state points on the coexistence line, under the most extreme quantum conditions studied herein ($\lambda_B^* = 1.9832$, $\gamma \cong 2.7 - 2.8$), $P = 30$ is sufficient to produce practical convergences in the centroid and in the instantaneous functions. For the solid state points at densities $\rho_N^* = 0.9$, 0.925, it is worthwhile to note that there is a slowing down of this convergence with decreasing temperatures ($\lambda_B^* = 0.2 \rightarrow 0.8$), which becomes more noticeable (a) for the triplet centroid quantities and (b) for the cI16 lattice, its openness playing a significant role in the fixing of the final particle distributions.

Table 2. Selected PIMC convergence features. Centroid (CM3) and instantaneous (ET3) first peaks (FP) in the close vicinities of the equilateral absolute maxima. Number in parentheses are one-standard deviation affecting the last digit(s) [1].

λ_B^*	ρ_N^*	$N_S \times P$	r_{FP-CM3}^*	g_{CM3}^{EQ}	r_{FP-ET3}^*	g_{ET3}^{EQ}
FLUID PHASE (fluid–FCC coexistence line)						
1.9832	0.348	864 × 30 864 × 40	1.5	19.7 (5) 19.9 (5)	1.4714	4.51 (2) 4.53 (3)
FCC PHASE (fluid–FCC coexistence line)						
0.8	0.573	864 × 12 864 × 18	1.3	55.6 (4) 56.3 (5)	1.3	15.4 (0) 15.4 (0)
FCC PHASE						
0.2	0.925	864 × 12 864 × 24	1.1	160.1 (7) 160.0 (15)	1.1286	93.4 (2) 93.1 (3)
0.8	0.9	864 × 12 864 × 24 864 × 36	1.1571	2339 (7) 3028 (16) 3177 (24)	1.1571	112.2 (1) 114.7 (2) 114.1 (2)
cI16 PHASE						
0.8	0.9	1024 × 12 1024 × 24 1024 × 36	1.1571	1469 (11) 1233 (10) 1334 (14)	1.1286	158.0 (2) 96.7 (2) 97.8 (3)

[1] 19.7(5) ≡ 19.7 ± 0.5; 160.0 (15) ≡ 160.0 ± 1.5; 2339(7) ≡ 2339 ± 7.

3.2. Closure Calculations

The current calculations at the actual fluid state points on the coexistence line use the new PIMC information obtained with sample sizes larger than those employed in [49]. (The new and the former results are in excellent agreement). The JF3 convolution integrals involve the h_2 extension to longer distances fixed with the fluid $S_{CM}^{(2)}(k)$ and $S_{ET}^{(2)}(k)$. The convolutions can be obtained by employing a well-known expansion in Legendre polynomials P_n [10,24]

$$\int dq_4 h_2(v_{14})h_2(v_{24})h_2(v_{34}) \cong \sum_{n=0}^{n_{max}} \pi(2n+1)P_n(\cos\phi)I_n(h_2,P_n), \tag{21a}$$

$$I_n(h_2,P_n) = \int_0^{y_{max}} dy\, y^2 h_2(y) f_n(y,s_{13}) f_n(y,u_{23}), \tag{21b}$$

$$f_n(y,z) = \int_{-1}^{+1} dx\, P_n(x) h_2\left(\sqrt{y^2 + z^2 - 2xyz}\right), \tag{21c}$$

where ϕ is the angle between s_{13} and u_{23}. The final JF3 results reported in Section 4 employ (a) $n_{max} = 30$ for the Legendre expansion; (b) $y_{max} = 20\sigma = 70$ Å (i.e., $y_{max}^* = 20$) as the maximum distance for h_2 data; and (c) trapezoidal quadratures with discretizations consisting of 2000 points for the y integrations and 1000 points for the x integrations. The latter parameters are sufficient to yield JF3 and AV3 results that can be compared with PIMC in a meaningful way. To grasp this point, some results at the highest-density fluid state point $\left(\rho_N^* = 0.789,\ \lambda_B^* = 0.2\right)$ will suffice. The JF3 centroid (CM3) and instantaneous (ET3) results in the close vicinities r_{FP}^* (first peaks FP) of their respective equilateral (EQ) absolute maxima, ($r_{FP} = 3.85$ Å, or $r_{FP}^* = 1.1$), behave as follows. (i) $n_{max} = 10$, $y_{max} = 50$ Å ($y_{max}^* \approx 14.3$), using 1000-point y integration, plus 500-point x integration leads to: $g_{CM3}^{EQ} = 42.930$, $g_{ET3}^{EQ} = 27.183$. (ii) $n_{max} = 10$, $y_{max} = 70$ Å ($y_{max}^* = 20$), using 2000-point y integration plus 1000-point x integration leads to $g_{CM3}^{EQ} = 42.931$, $g_{ET3}^{EQ} = 27.183$. (iii) $n_{max} = 30$, $y_{max} = 70$ Å ($y_{max}^* = 20$), using 2000-point y integration plus 1000-point x integration, lead to: $g_{CM3}^{EQ} = 42.932$, $g_{ET3}^{EQ} = 27.185$.

4. Results and Discussion

The results reported in this section are complemented with data in the Supplementary Materials.

4.1. The Pair Level Structures

Figure 1 shows representative pair radial correlation functions, centroid, and instantaneous, along the fluid–FCC solid coexistence (see also the Supplementary Materials for more information). The fluid functions (Figure 1a,b) display clear fluid-like features. Analogously, the FCC solid functions (Figure 1c,d) display the expected traits of FCC lattices. General comments on these pair radial functions are (i) the higher order in the solid functions that does not disappear with increasing distances; (ii) the outward shift and smoothing of the features with increasing quantum effects (on the coexistence line analyzed, one has $0.006 < \gamma < 2.81$); and (iii) the proximity between the locations of the fluid and solid first maxima (also between the dominant second-maximum regions), revealing that the system is ready to effect the change of phase. It is also interesting to note in passing that on the fluid side, the absolute maxima of the pair structures show dependences upon γ that can be fitted in the form $g_2(Max) = a\gamma^{-b}$, the associated linear correlation coefficients $r_{corr.}$ being reasonably good: (a) for the centroid functions, $a \cong 3.0042$, $b \cong 0.0687$, $r_{corr.} = -0.9982$; and (b) for the instantaneous functions, $a \cong 1.8863$, $b \cong 0.1233$, $r_{corr.} = -0.9999$. Furthermore, the concordance at the pair level between the results in the $\{r\}$ and the $\{k\}$ spaces is excellent. The fluid radial functions are fully consistent in particular with the configurational maximal values arising from Equations (14) and (15): the fluid phase maximal values obtained remain $S_2^{(C)} < 0.1$. Moreover, Table 3 contains the observed variations in the maximal values of $S_2^{(C)}$ corresponding to the FCC centroid and instantaneous cases. For the current calculations, a representative FCC-set of maximal wave vectors can be defined by their k-integer components: $\{(-6,6,6), (-6,6,-6,), (6,6,6)\}$.

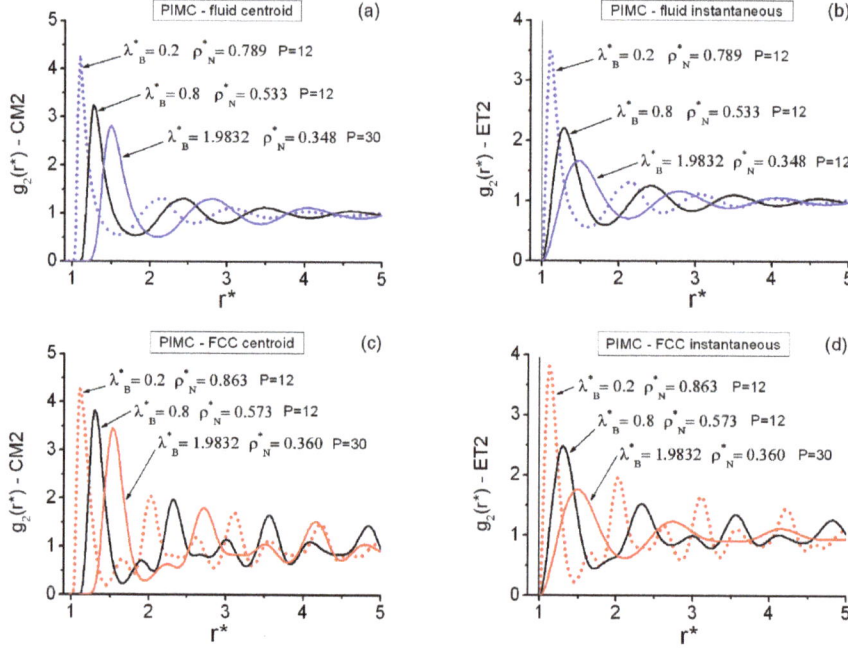

Figure 1. PIMC pair radial correlation functions along the quantum hard spheres (QHS) fluid–FCC (face-centered cubic) solid coexistence line (six state points in Table 1). The arrangement should be clear according to the (ρ_N^*, λ_B^*) values shown. (**a**) Fluid centroid functions. (**b**) Fluid instantaneous functions. (**c**) FCC centroid functions. (**d**) FCC instantaneous functions. The vertical line at $r^* = 1$ in (**b**,**d**) marks the position of the hard core.

Table 3. Solid phase variations in the maximal values of the centroid (CM2) Equation (14) and instantaneous (ET2) Equation (15) configurational structure factors at the pair level fixed with PIMC.

		FCC PHASE on the Coexistence Line		
λ_B^*	ρ_N^*	$N_S \times P$	$S_{CM2}^{(C)}(k_{max})$	$S_{ET2}^{(C)}(k_{max})$
0.2	0.863	864 × 12	0.803–0.764	0.786–0.748
0.4	0.731	864 × 12	0.791–0.751	0.738–0.702
0.6	0.635	864 × 12	0.778–0.738	0.686–0.649
0.8	0.573	864 × 12	0.784–0.752	0.643–0.613
1.2543	0.465	864 × 24	0.771–0.732	0.532–0.503
1.9832	0.360	864 × 30	0.743–0.691	0.393–0.356
		FCC PHASE		
0.2	0.925	864 × 12	0.886–0.866	0.867–0.849
		864 × 24	0.883–0.865	0.864–0.847
0.8	0.9	864 × 12	0.986–0.984	0.898–0.894
		864 × 24	0.989–0.987	0.902–0.898
		864 × 36	0.989–0.988	0.901–0.900
		cI16 PHASE		
0.2	0.925	1024 × 12	0.726–0.698	0.710–0.682
		1024 × 24	0.732–0.705	0.717–0.689
0.8	0.9	1024 × 12	0.793–0.784	0.741–0.732
		1024 × 24	0.781–0.774	0.712–0.705
		1024 × 36	0.777–0.771	0.708–0.702

Figure 2 shows the pair radial correlation functions, centroid and instantaneous, of the FCC and cI16 state points at the moderately high densities $\rho_N^* = 0.9$, 0.925. There is a sharp contrast between the FCC and cI16 structures, since the usual coordination shells existing in the highly regular FCC lattice are absent from cI16. The most characteristic trait of cI16 is, perhaps, the presence of a convoluted inner structure, with two conspicuous big dips, for distances below $r^* \approx 2.5$. The FCC solid structures (Figure 2a,c) are the "compression" (at constant temperature) of the corresponding FCC structures on the coexistence line. The current cI16 results (Figure 2b,c) agree feature for feature with the pair structures displayed by the bcc-qIII phases in [67]. (Differences between the first peaks are due to the B-spline smoothing carried out in Figure 9 of [67]; see the Suppplementary Material for non-smoothed data). This deserves to be highlighted, since the PIMC-QHS origins of both types of structures are not related: the former bcc-qIII phases arose from the evolution of initially perfect BCC lattices ($N_S = 432$), while the present (delocalized) cI16 phases are just the results obtained from the evolution of initially perfect cI16 lattices ($N_S = 1024$). To complete the foregoing information, Table 3 also contains the variations in the maximal values of the respective cI16-configurational $S_2^{(C)}$ structure factors. They are consistent with the behavior reported in [67]. For the current calculations, a representative cI16-set of maximal wave vectors can be defined by their k-integer components $\{(-8,8,0), (0,8,-8,), (8,8,0)\}$.

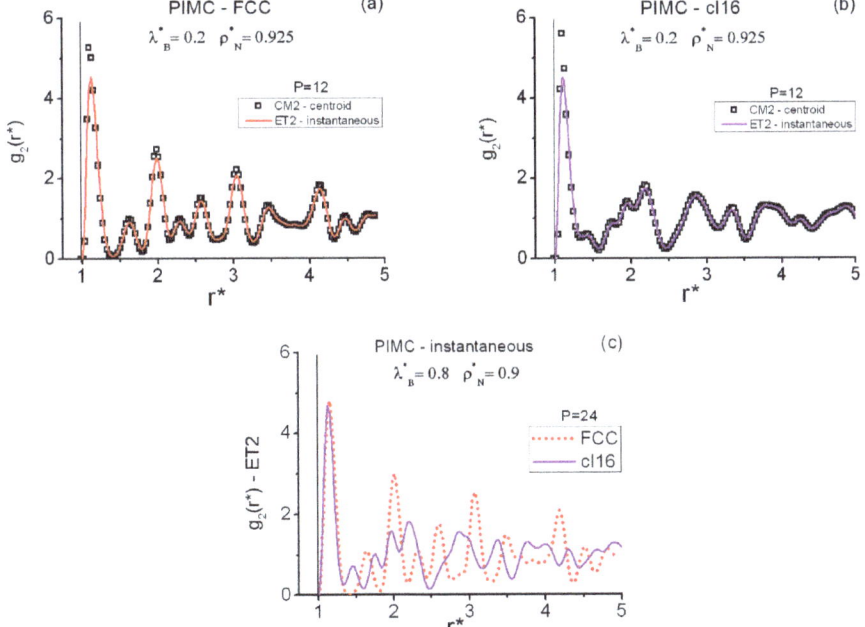

Figure 2. PIMC pair radial correlation functions in the region of moderately high densities for the cubic-based QHS solid phases FCC and cI16. No smoothing of the simulation results has been carried out. The vertical line at $r^* = 1$ in (**a**–**c**), marks the position of the hard core.

There is still the further question related to the characterization of cI16 phases via the fractional displacement parameter x. In the quantum case, the delocalization makes this task a three-fold one, since there are three types of distinct structures. Given the current scope, only the centroid and instantaneous x estimates are determined in this work. A convenient way is through the tabulation for perfect cI16 lattices of $\left(x, S_2^{(C)}(k_{\max})\right)$, which can be computationally fixed by varying x. Thus, for the interval $0.025 \leq x \leq 0.038$, using $\Delta x = 0.001$, the related parabolic least-squares fitting (better than just linear) leads to

$$S_2^{(C)}(k_{max}) = 1.0931 - 8.269x - 108.75x^2; \text{ CM2 and ET2}, \tag{22}$$

which guarantees absolute errors of orders $\leq 10^{-4}$ in the estimated values of the reference maximal structure factors (for higher precision, the reader is referred to the tabulation in the Supplementary Materials). Note that the higher the x is, the lower the $S_2^{(C)}(k_{max})$ becomes, as expected. In this regard, note that for $x = 0$, which is out of the above interval, one must retrieve the perfect BCC limit $S_2^{(C)}(k_{max}) = 1$. Consideration of the actual calculated values of the maximal $S_{CM2}^{(C)}(k_{max})$ and $S_{ET2}^{(C)}(k_{max})$ in Table 3 yields the cI16 variations: (i) at $(\rho_N^* = 0.925, \lambda_B^* = 0.2; P = 12)$, $0.0314 \leq x \leq 0.0332$ for CM2, and $0.0325 \leq x \leq 0.0343$ for ET2; and (ii) at $(\rho_N^* = 0.9, \lambda_B^* = 0.8; P = 24)$, $0.0277 \leq x \leq 0.0282$ for CM2, and $0.0323 \leq x \leq 0.0328$ for ET2. These values show the expected behavior: (a) they are larger for the instantaneous structures; (b) they are consistent with cI16 values reported in the literature [76,77,80]; and (c) the CM2–ET2 differences increase with the quantum effects. Another point to consider here is related to the fact that samples of classical hard spheres can be "squeezed" more than samples of quantum hard spheres, because of the latter's "preemptive" repulsions. This means that via low temperatures, one can expect QHS–cI16 phases to appear for lower densities than in the classical hard-sphere system $(\rho_N^* \geq 1.1)$ [80], which is indeed the case.

4.2. Triplets in the Fluid Phase

Figures 3–5 show the main features of the fluid triplet correlations analyzed in this work. Several general trends can be easily identified in Figure 3, which collects results at two state points along the lowest isotherm $\lambda_B^* = 1.9832$. First, as occurred on the pair level, the centroid CM3 features are far more pronounced than those of the instantaneous ET3 case. Second, and associated with the equilateral data, one notes that the first maximum and the first minimum positions of a given $g_3(r^*, r^*, r^*)$ occur in the close vicinities of the corresponding first maximum and first minimum of the associated $g_2(r^*)$ shown in Figure 1. Third, although the closures KS3 and JF3 fail to reproduce the exact PIMC behavior, their average AV3 shows a remarkable performance for both the centroid and the instantaneous correlations. Fourth, as the density increases along isotherms, and when going toward longer distances, AV3 loses predictive power to fit the profiles of the isosceles correlations $g_3(r^*, s^*, s^*)$. In relation to this, see Figure 3d, where $s^* = s_M^*$ is such that $g_3(s_M^*, s_M^*, s_M^*) \cong$ absolute equilateral maximum.

Finer equilateral facts follow. (i) Figure 3a,b displays explicitly, at state point $(\rho_N^* = 0.1, \lambda_B^* = 1.9832)$, the equilateral asymptotic behavior $g_3(r^*, r^*, r^*) \to 1$ with increasing r^* for the PIMC centroid and instantaneous correlations. (ii) Figure 3c illustrates the isosceles asymptotic behavior $g_3(r^*, s^*, s^*) \to g_2(r^*)$, when the two s^* distances increase. (iii) As seen, the short-range behavior of AV3 is non-correct (due to that of JF3), whereas KS3 behaves properly. (iv) At constant temperature, there is a sharpening and shifting inwards of the structures with increasing density. For example, at $\lambda_B^* = 1.9832$ in the vicinities (r_{FP}^*) of the equilateral first maxima, the $g_3^{EQ} = g_3(r^*, r^*, r^*)$ behave as follows: (1) $\rho_N^* = 0.1$, $(r_{FP}^* = 1.9, g_{CM3}^{EQ} = 2.16)$ and $(r_{FP}^* = 2, g_{ET3}^{EQ} = 1.41)$; (2) $\rho_N^* = 0.3$, $(r_{FP}^* = 1.5571, g_{CM3}^{EQ} = 12.65)$ and $(r_{FP}^* = 1.5429, g_{ET3}^{EQ} = 3.54)$; and (3) $\rho_N^* = 0.348$, $(r_{FP}^* = 1.5, g_{CM3}^{EQ} = 19.74)$ and $(r_{FP}^* = 1.4714, g_{ET3}^{EQ} = 4.51)$. An analogous behavior can be observed at the pair level. (Use $\sigma = 3.5$ Å and rounding-off to two decimal places to transform the foregoing r^* into the actual r of the (M, σ) system utilized in the current calculations, e.g., $r^* = 1.5571 \to r^* = 5.5$ Å).

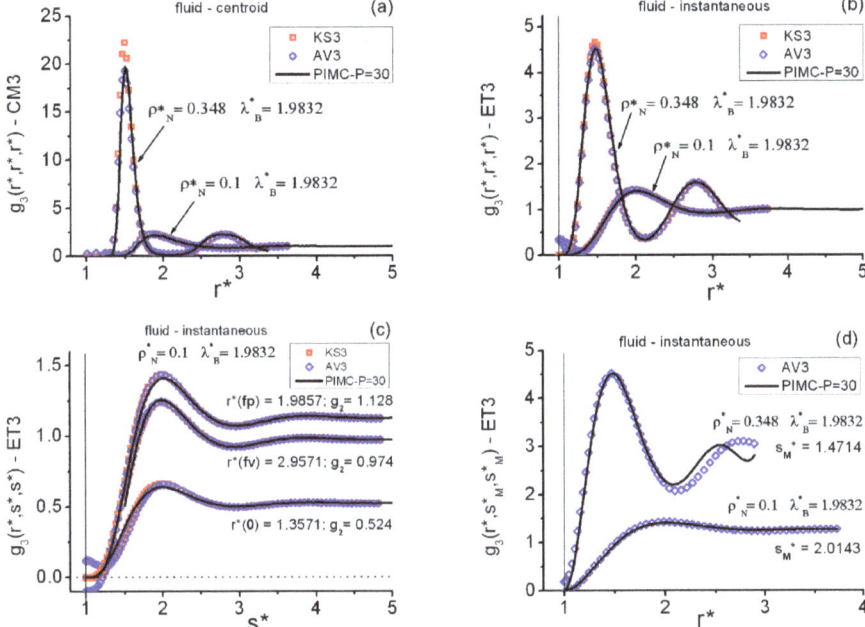

Figure 3. Typical behaviors of the centroid and instantaneous triplet correlations in the QH fluid at two representative state points. KS3 = Kirkwood superposition, Equation (17); AV3 = average closure, Equation (19); PIMC = path integral Monte Carlo. (**a**) Centroid equilateral; (**b**) instantaneous equilateral; (**c**) instantaneous isosceles, with pair $g_2(r^*)$ asymptotic values shown (increasing s^*) at three selected r^* (close to the pair first maximum fp, close to the pair first minimum fv, and with 0 being a pair close-range distance); (**d**) r^* profiles of the heights in the close vicinity of the first maxima of the instantaneous isosceles correlations (s_M^* = distance in the close vicinity of where the absolute equilateral maximum appears). The vertical line at $r^* = 1$ in (**b**–**d**) marks the position of the hard core.

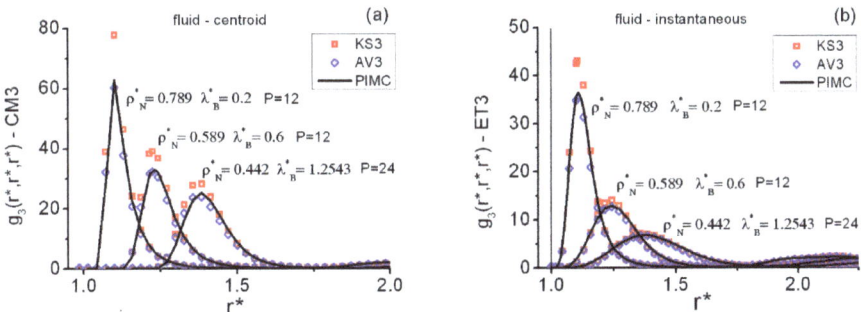

Figure 4. Typical forms of the centroid and the instantaneous equilateral correlations in the QHS fluid at three representative state points on the fluid–FCC coexistence line. Acronyms for methods as in Figure 3. (**a**) Fluid centroid functions. (**b**) Fluid instantaneous functions. The vertical line at $r^* = 1$ in (**b**) marks the position of the hard core.

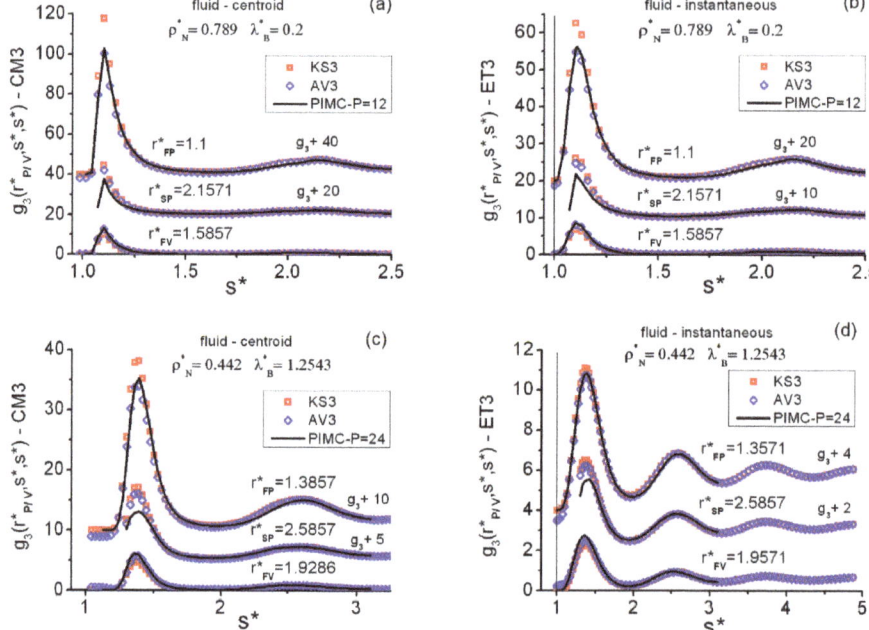

Figure 5. Typical behaviors of the centroid and instantaneous isosceles correlations in the QHS fluid at two representative state points on the fluid–FCC coexistence line. Acronyms for methods as in Figure 3. Results at three especial r^* distances of the equilateral correlations very close to the respective: first maximum (FP), first minimum (FV), and second maximum (SP). (**a**) Upper plots shifted by +20 and +40. (**b**) Upper plots shifted by +10 and +20. (**c**) Upper plots shifted by +5 and +10. (**d**) Upper plots shifted by +2 and +4. The vertical line at $r^* = 1$ in (**b**,**d**) marks the position of the hard core.

Figure 4 shows the equilateral correlations at three representative state points on the fluid side of the coexistence line. The aforementioned trends of KS3 and AV3, as compared to PIMC, appear again for both types of correlations CM3 (Figure 4a) and ET3 (Figure 4b). In going from higher to lower densities/temperatures on the fluid side, one observes that the larger the quantum effects, the flatter the structural triplet features become.

Table 4 contains the absolute maxima, fixed with quadratic interpolations of the adequate PIMC data, of the fluid equilateral correlations. (See the Supplementary Materials for more related numerical data). Once more, in an attempt to connect the foregoing maximum behaviors with the quantum parameter γ, one notes that simple empirical decay fittings $g_3^{EQ}(Max) = a\gamma^{-b}$ can be found for the centroid and for the instantaneous cases, their associated linear correlation coefficients $r_{corr.}$ being reasonably good. Thus, one finds for the centroid case $a \cong 23.6702, b \cong 0.1877, r_{corr.} = -0.9959$ and for the instantaneous case $a \cong 6.437, b \cong 0.3449, r_{corr.} = -0.9999$. This general pattern is to be regarded as a reflection of the very same observed at the pair level. Three additional points are worthwhile to mention: (i) the quality of this type of fitting remains comparable if one tries the modification $g_3^{EQ}(Max) = a\gamma^{-b} + c$; (ii) exponential decays, e.g., $g_3^{EQ}(Max) = a\exp(-b\gamma)$, give poor fittings; and (iii) the potential energy discontinuity at $r^* = 1$ precludes one from retrieving the classical limit at $\lambda_B^* = 0$. Although there is no apparent physical basis for the empirical γ pattern found, this line of thought might be well worth exploring in future work.

Table 4. Absolute maxima of the PIMC equilateral correlations $g_3^{EQ} = g_3(r^*, r^*, r^*)$ on the fluid and the FCC sides of the QHS coexistence line. Discretizations at $\lambda_B^* = 0.8$ and 1.9832: $P = 12$ and 30, respectively. $r^* = r/\sigma$. Four decimals shown in g_3^{EQ} to avoid rounding-off errors.

	FLUID–CENTROID–			FCC–CENTROID–		
λ_B^*	ρ_N^*	r_{Max}^*	g_{CM3}^{EQ}	ρ_N^*	r_{Max}^*	g_{CM3}^{EQ}
0.2	0.789	1.1029	63.1089	0.863	1.1097	87.4597
0.4	0.672	1.1690	42.9419	0.731	1.1867	66.8718
0.6	0.589	1.2313	32.7749	0.635	1.2504	56.1959
0.8	0.533	1.2841	29.4898	0.573	1.3042	55.7136
1.2543	0.442	1.3841	25.1954	0.465	1.4051	48.7081
1.9832	0.348	1.5096	19.8989	0.360	1.5360	40.1005
	FLUID–INSTANTANEOUS–			FCC–INSTANTANEOUS–		
λ_B^*	ρ_N^*	r_{Max}^*	g_{ET3}^{EQ}	ρ_N^*	r_{Max}^*	g_{ET3}^{EQ}
0.2	0.789	1.1101	36.7840	0.863	1.1255	55.7274
0.4	0.672	1.1832	19.2805	0.731	1.1990	31.3301
0.6	0.589	1.2419	12.9259	0.635	1.2584	20.1514
0.8	0.533	1.2890	10.0773	0.573	1.3029	15.4292
1.2543	0.442	1.3752	6.8439	0.465	1.3880	9.1774
1.9832	0.348	1.4820	4.5227	0.360	1.4947	5.4411

Figure 5 contains a quick description of the isosceles correlations $g_3(r^*, s^*, s^*)$ at two representative fluid state points, for the centroids CM3 in panels (a)–(c) and for the instantaneous ET3 in panels (b)–(d). Three especial r^* distances are selected from the $g_3(r^*, r^*, r^*)$ information obtained at each state point, namely r_{FP}^*, r_{FV}^*, and r_{SP}^*, which are positions in the close vicinities of the equilateral maxima and minima: first maximum (FP), first minimum (FV), and second maximum (SP), respectively. Apart from the expected AV3 unphysical behavior for $r^* \leq 1$, the good overall performance of AV3 is certainly surprising. Two weak points are to be remarked. First, the AV3 (and KS3) behavior for low s^* distances, $1 < s^* < 1.5$, when r^* increases: for example, at r_{SP}^* where the closure maxima are overestimated. (This is directly related to the AV3 trend displayed by the upper profile plot in Figure 3d). Second, Figure 6 shows a detailed image of the isosceles deterioration of the PIMC–AV3 agreement with increasing densities, the worse results for centroids being a consequence of this key fact (centroids mimic a fluid at a higher density than the actual one).

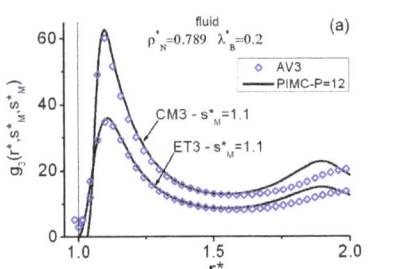 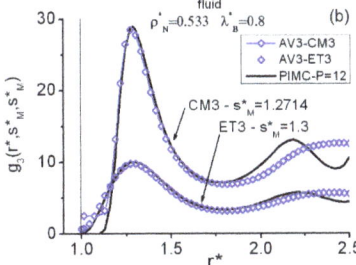

Figure 6. QHS fluid centroid (CM3) and instantaneous (ET3) r^* profiles of the heights in the close vicinities of the first peaks of the isosceles correlations at two selected state points on the fluid–FCC coexistence line. (**a**) Fluid functions at ($\rho_N^* = 0.789$, $\lambda_B^* = 0.2$). (**b**) Fluid functions at ($\rho_N^* = 0.533$, $\lambda_B^* = 0.8$). $s_M^* =$ distance in the close vicinity of the absolute maximum of the equilateral correlations. Acronyms for methods as in Figure 3. The vertical line at $r^* = 1$ marks the position of the hard core.

4.3. FCC triplets on the Fluid–Solid Coexistence Line

Table 4 and Figures 7 and 8 show selected results for the PIMC equilateral and isosceles correlations of FCC state points on the solid side of the fluid–FCC coexistence line, within the approximations obtainable from Equations (7) and (8). For visualization purposes, the associated PIMC fluid results are also incorporated into these figures.

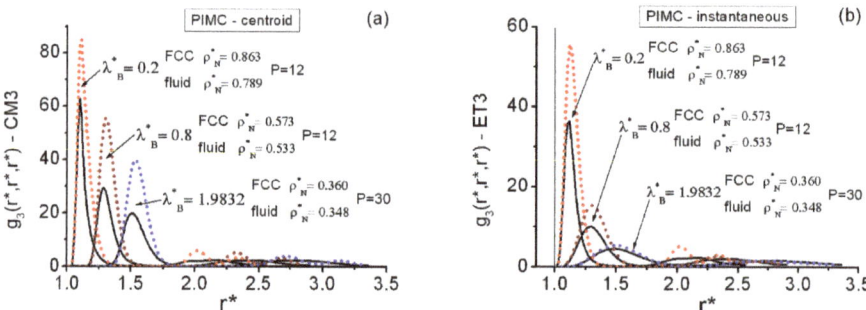

Figure 7. Comparison between PIMC equilateral structures on both sides of the fluid–FCC coexistence line at selected state points. (**a**) Centroid functions. (**b**) Instantaneous functions. The vertical line at $r^* = 1$ in (**b**) marks the position of the hard core.

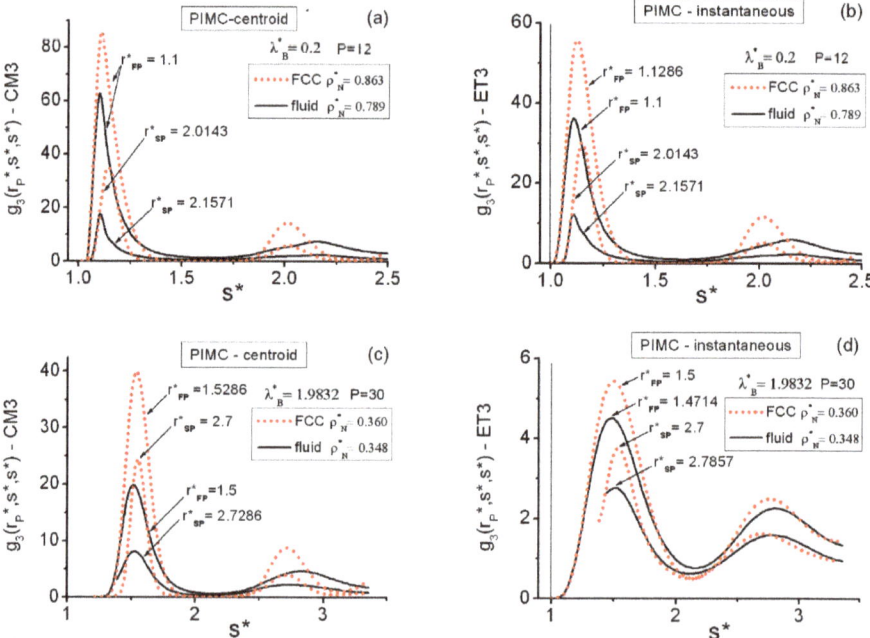

Figure 8. Comparison of PIMC isosceles triplet structures on both sides of the fluid–FCC coexistence line at selected state points and r_{FP}^* and r_{SP}^* slices. These r^* are very close to the first (FP) and second (SP) maxima of the corresponding equilateral structures. (**a**) Centroid functions at $\lambda_B^* = 0.2$. (**b**) Instantaneous functions at $\lambda_B^* = 0.2$. (**c**) Centroid functions at $\lambda_B^* = 1.9832$. (**d**) Instantaneous functions at $\lambda_B^* = 1.9832$. The vertical line at $r^* = 1$ in (**b**,**d**) marks the position of the hard core.

116

A general idea can be obtained by observing Table 4. The fluid and solid absolute maximum positions are close to one another, and the structures are shifted outwards with increasing quantum effects. Moreover, higher g_3 values appear on the solid side (e.g., at $\lambda_B^* = 0.2$, $\approx +39\%$ for CM3, and $\approx +51\%$ for ET3). This trend is far more pronounced for the centroid correlations, the ratio increasing monotonically with λ_B^* (e.g., at $\lambda_B^* = 1.9832$, $\approx +102\%$ for CM3). However, for the instantaneous ET3 correlations, such a ratio is not monotonic; it goes through a maximum (at $\lambda_B^* = 0.4$, $\approx +62\%$) and then falls monotonically (at $\lambda_B^* = 1.9832$, $\approx +20\%$). These behaviors can be ascribed to the two effects present on the coexistence line. On the one hand, there is the decreasing density, which contributes to diminishing the structural features. On the other hand, there is the increasing delocalization with λ_B^*, which makes PI structures become more and more smeared out, the instantaneous case being always much more sensitive to this. As regards the question of finding a γ–fitting of the solid equilateral absolute maxima, the situation is less clear than on the fluid side (γ is slightly higher on the solid side). Although one can obtain reasonable dependences $g_3^{EQ}(Max) = a\gamma^{-b}$ ($r_{corr.} < -0.991$), on closer inspection, these fittings cannot cope with the apparent inflection in $0.13 < \gamma < 0.3$ (or in $0.6 < \lambda_B^* < 0.8$) for centroids $g_{CM3}^{EQ}(Max)$, nor with the large discrepancies for low γ between the original and the estimated instantaneous values $g_{ET3}^{EQ}(Max)$.

In Figure 7, one observes that the equilateral FCC and fluid $g_3(r^*,r^*,r^*)$ patterns are qualitatively similar within the first maximum regions. It is also noticeable that the FCC state points develop easily identifiable peak structures with increasing distances ($r^* \gtrsim 2$). The main two maxima of the FCC equilateral triplets can be put into direct correspondence with the main two maxima obtained at the FCC pair level (Figure 1), since they appear located close to one another.

The FCC $g_3(r^*,r^*,r^*)$ display deep first valleys, almost at the zero-ground level, appearing for both the centroid and the instantaneous structures, e.g., for centroids and $\rho_N^* = 0.573$, the region in Figure 7 located in $1.6 \lesssim r^* \lesssim 2.1$. In general, this feature is far more pronounced in the centroid structures than in the instantaneous structures and is consistently shifted outwards with increasing quantum effects. If comparison with Figure 1c,d is made, one notes that this triplet region corresponds to the FCC pair region where the smallest maximum shows up. (Such region fades away with increasing quantum effects in the instantaneous case, Figure 1d). To get a feeling of the depth of these valleys, it seems worthwhile to quote some significant results: (a) at $\left(\rho_N^* = 0.863, \lambda_B^* = 0.2\right)$, within the range $1.4143 \leq r^* \leq 1.8143$, the equilateral centroid and instantaneous values remain $g_{CM3}^{EQ} \lesssim 0.07$ and $g_{ET3}^{EQ} \lesssim 0.09$, respectively; (b) at $\left(\rho_N^* = 0.573, \lambda_B^* = 0.8\right)$, within the range $1.6143 \leq r^* \leq 2.0714$, the equilateral centroid values remain $g_{CM3}^{EQ} \lesssim 0.1$, whereas the equilateral instantaneous values reach the same upper bound $g_{ET3}^{EQ} \lesssim 0.1$ within the narrower range $1.6714 \leq r^* \leq 1.9571$; and (c) at $\left(\rho_N^* = 0.360, \lambda_B^* = 1.9832\right)$, within the range $1.9 \leq r^* \leq 2.3571$, the equilateral centroid values remain $g_{CM3}^{EQ} \lesssim 0.08$, whereas the equilateral instantaneous values g_{ET3}^{EQ} do not go below 0.15 within their related first valley. (See the Supplementary Materials for more data on the coexistence line).

In addition, Figure 8 contains typical isosceles $g_3(r^*,s^*,s^*)$ behaviors of the fluid and the FCC solid at the lowest ($\lambda_B^* = 0.2$) and the highest ($\lambda_B^* = 1.9832$) de Broglie wavelengths. These graphs display significant r^*–slices (i.e., at r_{FP}^* and r_{SP}^*) of the tabulated functions in the close vicinities of the corresponding first (FP) and second (SP) maxima of the equilateral correlations. The parallels between the triplets of the solid and fluid phases coexisting at equilibrium are manifest once more.

4.4. Triplets in the FCC and cI16 Denser Solid Structures

Figure 9 and Table 5 contain equilateral PIMC results for the FCC and cI16 state points in the region of moderately high densities ($\rho_N^* = 0.9, 0.925$). The centroid CM3 and the instantaneous ET3 correlation results, with $P = 12$ for both lattices at ($\rho_N^* = 0.925$, $\lambda_B^* = 0.2$), are P converged (Table 2). At ($\rho_N^* = 0.9$, $\lambda_B^* = 0.8$), convergence for the instantaneous correlations with $P = 36$ is guaranteed (practical convergence already occurs with $P = 24$), whereas for the centroid correlations, there is still room for further improvement. Nevertheless, the centroid results obtained with $P = 36$ are expected to capture well the related global features. This contrasts with the more rapid P convergence for centroids at the pair level.

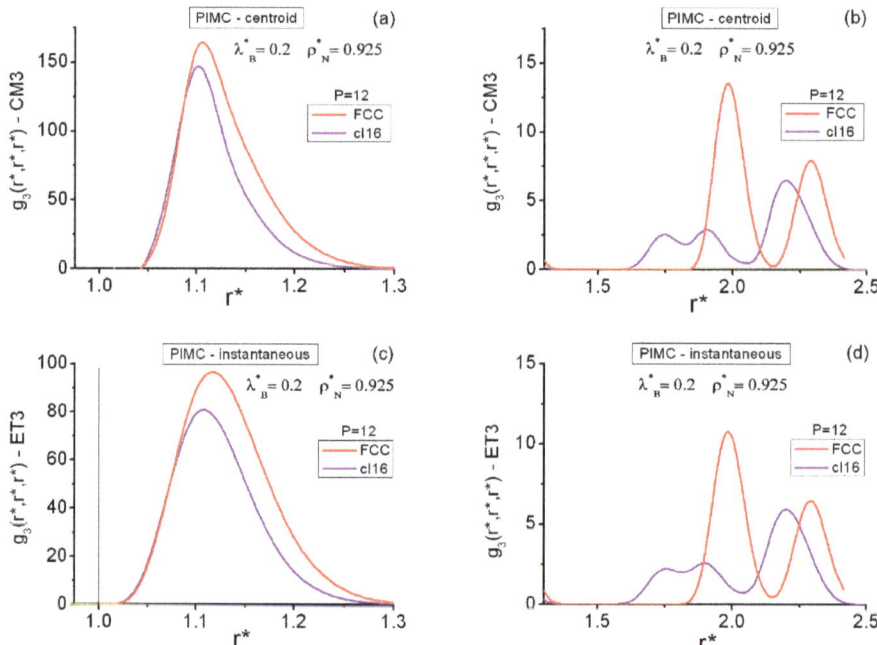

Figure 9. PIMC results for the FCC and cI16 equilateral structures in the region of moderately high densities ρ_N^*. The graphs are split horizontally into two parts to avoid the flat g_3 range of distances and show the secondary maximum regions on a visible scale. (**a**) Centroid functions in the short-distance range. (**b**) Centroid functions in the medium-distance range. (**c**) Instantaneous functions in the short-distance range. (**d**) Instantaneous functions in the medium-distance range. The vertical line at $r^* = 1$ in (**c**) marks the position of the hard core.

Table 5. Average salient features of the cI16 and FCC equilateral centroid (CM3) and instantaneous (ET3) correlations $g_3^{EQ} = g_3(r^*, r^*, r^*)$. PIMC results in the close vicinities of the maxima and minima. Maxima: first FP, second SP, third TP, fourth F4P. Minima: first FV, second SV, third TV.

		(ρ_N^*=0.925, λ_B^*=0.2)		
	cI16–(r^*,g_3^{EQ})–PIMC–P=12		FCC–(r^*,g_3^{EQ})–PIMC–P=12	
	CM3	ET3	CM3	ET3
FP	(1.1, 146.80)	(1.1, 79.46)	(1.1, 160.07)	(1.1286, 93.38)
FV	(1.4714, 0)	(1.4714, 4×10^{-5})	(1.5571, 0)	(1.5571, 0)
SP	(1.7571, 2.52)	(1.7571, 2.25)	(1.9857, 13.53)	(1.9857, 10.80)
SV	(1.8143, 1.87)	(1.8143, 2.02)	(2.1571, 0.31)	(2.1571, 0.51)
TP	(1.9, 2.91)	(1.9, 2.61)	(2.3, 7.82)	(2.3, 6.42)
TV	(2.0429, 0.51)	(2.0429, 0.75)		
F4P	(2.1857, 6.40)	(2.1857, 5.86)		

Table 5. Cont.

	(ρ_N^*=0.9, λ_B^*=0.8)			
	cI16-(r^*,g_3^{EQ})–PIMC-P=36		FCC-(r^*,g_3^{EQ})–PIMC-P=36	
	CM3	ET3	CM3	ET3
FP	(1.1571, 1334)	(1.1286, 97.82)	(1.1571, 3177)	(1.1571, 114.08)
FV	(1.4429, 0)	(1.4571, 0)	(1.5857, 0)	(1.5857, 0)
SP	(1.7571, 83.72)	(1.7571, 4)	(2.0143, 425.91)	(2.0143, 18.70)
SV	(1.8714, 0)	(1.8429, 0.92)	(2.1429, 0)	(2.1714, 0.245)
TP	(1.9571, 72.32)	(1.9571, 3.12)	(2.3286, 260.89)	(2.3286, 11.90)
TV	(2.1, 0)	(2.0714, 0.43)		
F4P [1]	(2.3, 0.44)	(2.2143, 5.68)		

[1] There is a cI16 small bump at $r^* = 2.2143$, $g_{CM3}^{EQ} = 0.01$ ($P = 12$), 0.18 ($P = 24$), 0.14 ($P = 36$).

In Figure 9, the equilateral correlations of the FCC and cI16 state points at ($\rho_N^* = 0.925$, $\lambda_B^* = 0.2$) are considered within $r^* < 2.5$. Three well-defined features can be seen in each case, and they can be put into correspondence with the results obtained at the related distances on the pair level (Figure 2). Thus, three separated maxima arise from the triplet FCC calculations (as occurred on the coexistence line). However, four maxima arise from the triplet cI16 calculations, with the second and third forming an overlapping structure. This reminds one of the characteristic shallow split showing up past the first maximum in the $g_2(r^*)$ of amorphous systems [99]. Moreover, the FCC features are more pronounced than those of cI16, as was to be expected. In addition, for $1.5 < r^* < 2.5$, cI16 and FCC are somewhat complementary regarding the positions of their peaks. One observes the clear quantitative differences between the centroid CM3 and the instantaneous ET3 results. The patterns of the salient features shown in Table 5 for the two density–temperature conditions are fully consistent with each other and with the underlying pair information (Figure 2).

A closer inspection of the equilateral flat regions between the first and the second maxima may be worth carrying out. The following results correspond to the discretizations: (i) $P = 24$ at ($\rho_N^* = 0.925$, $\lambda_B^* = 0.2$), although $P = 12$ results are not significantly different; and (ii) $P = 36$ at ($\rho_N^* = 0.9$, $\lambda_B^* = 0.8$).

(a) As regards the FCC results, these regions are related to those found on the coexistence line, but now the behavior is much more extreme: the zero-ground level is effectively reached. At ($\rho_N^* = 0.925$, $\lambda_B^* = 0.2$), centroid values $g_{CM3}^{EQ} \equiv 0$ are obtained within $1.4714 \leq r^* \leq 1.6429$, while instantaneous values $g_{ET3}^{EQ} \equiv 0$ are within $1.5286 \leq r^* \leq 1.5857$. Moreover, at ($\rho_N^* = 0.9$, $\lambda_B^* = 0.8$), centroid values $g_{CM3}^{EQ} \equiv 0$ are obtained within $1.2714 \leq r^* \leq 1.9$, while instantaneous values $g_{ET3}^{EQ} \equiv 0$ are within $1.4714 \leq r^* \leq 1.7$.

(b) The situation of cI16 is less severe, although with increasing quantum effects, some of the previous traits also arise. Thus, at ($\rho_N^* = 0.925$, $\lambda_B^* = 0.2$), centroid values remain $g_{CM3}^{EQ} \lesssim 0.04$ within $1.3286 \leq r^* \leq 1.5857$, with $g_{CM3}^{EQ} \equiv 0$ only for $1.44 \leq r^* \leq 1.47$, while the instantaneous values are above zero in that latter region ($0 < g_{ET3}^{EQ} \lesssim 0.08$). At ($\rho_N^* = 0.9$, $\lambda_B^* = 0.8$), centroid values $g_{CM3}^{EQ} \equiv 0$ appear within $1.2429 \leq r^* \leq 1.6429$, while instantaneous values $g_{ET3}^{EQ} \equiv 0$ do only for $1.44 \leq r^* \leq 1.47$.

The solid triplet flat regions arise from the combination of the role of the QHS interactions and the unavailability of space due to the variations in ρ_N^* and λ_B^*. As a result, the solid equilateral structures analyzed turn out to be simpler than their pair radial counterparts (Figure 2), which is especially true of the cI16 lattice. Use of this fact might find application to characterizing irregular solid structures and/or monitoring their formation. (See the Supplementary Materials for more information on these structures). Another observation is related to the order shown by these two lattices. FCC appears as more ordered than *cI16*, which is clear from the respective regularities in their pair and triplet correlation functions and also from the maximal configurational structure factors (Table 3). Therefore, under the same (ρ_N^*, λ_B^*) conditions, FCC entropies (and free energies) [67] must be lower than those of *cI16*, the determination of these properties being possible via the Einstein crystal quantum technique [66,67].

5. Conclusions

This article has analyzed several real space triplet correlation issues in the PI–QHS system under conditions in which quantum exchange can be neglected. Triplet PI centroid and instantaneous correlations (equilateral and isosceles) in significant fluid and FCC–solid-state points have been studied. Furthermore, the positive identification of the formerly denoted bcc–qIII solid phases [67,78] with proper quantum cI16 solid phases has been achieved by utilizing information at the pair level (radial structures and maximal structure factor values). Triplet calculations have also been carried out at two cI16 state points. The results lead to the following conclusions.

(1) Fluid phase and the use of closures.

   (a) The centroid results display far more structured triplet functions than the instantaneous results. These structures tend to be shifted outwards with increasing λ_B^* (delocalization) and inwards with increasing ρ_N^* (localization).

   (b) From the comparison between PIMC and the closure results, one concludes that the role of pair correlations in shaping triplet structures is more relevant in the quantum domain than previously thought. The combined use of KS3 (for short range) plus AV3 = (KS3+JF3)/2 (beyond short range), although not exact, is found to be a useful and simple choice to understand the related main {r} triplet features, either centroid or instantaneous, of fluids with quantum diffraction effects.

   (c) The AV3 success appears to be linked with the fact that this closure adopts the form of a "complete" h_2 expansion truncated to first order in the density, which includes explicitly the triple-h_2 product absent from JF3. Given that along isotherms, AV3 deteriorates with increasing distances as the fluid–solid coexistence is approached, improvements on AV3 may be of interest and should incorporate at least second-order density terms in the h_2 expansion.

   (d) The foregoing finding extends the previous results obtained in [17] for liquid para-hydrogen and liquid neon, since the current study has involved a purely repulsive interparticle potential. Therefore, applications of an improved AV3 (supplemented with KS3 as said above) might be expected to provide reliable pictures of what is behind triplet correlations in fluid helium over a wide range of conditions [4,15].

(2) The fluid–FCC solid coexistence line.

   (a) There is a close correspondence between the positions of the main structural features at short range of both phases at equilibrium, not only at the pair level but also at the triplet level. Such a phase correspondence between triplet positions appears in both the equilateral and the isosceles correlations. These are clear signs of the system readiness to undergo the phase transition.

   (b) The triplet features are far more pronounced on the solid side. In addition, the centroid features are always sharper than those of their instantaneous counterparts (e.g., more elevated peak regions and lower valley regions for centroids).

(3) On the fluid side, the absolute maxima of the pair and the triplet-equilateral correlations, centroid and instantaneous, appear to follow empirical behaviors that depend on the quantum parameter $\gamma = \rho_N^* \lambda_B^{*3}$ in the general form $g_3^{EQ}(\text{Max}) = a\gamma^{-b}$. For systems in which repulsive particle

interactions dominate, this might be a further structural signature of the fluid phase on the quantum crystallization line [17,18,49], and it deserves further examination.

(4) FCC and cI16 solid phases.

  (a) FCC state points show a significantly higher order than their cI16 counterparts, at the same density–temperature conditions, which can be ascribed to the openness of the cI16 lattice. This is observed for the two structures, centroid and instantaneous, in all the forms computed $\left(g_2, g_3, S_2^{(C)}\right)$. Roughly speaking, at a given state point, using the maximal values of the configurational structure factors, one finds that $S_{CM2}^{(C)}(FCC)/S_{CM2}^{(C)}(cI16) \approx S_{ET2}^{(C)}(FCC)/S_{ET2}^{(C)}(cI16)$. FCC entropies must certainly be lower than their cI16 counterparts, and it is tempting to explore the relationships between the solid entropy and the values of the quantum structure factors in future work.

  (b) Within the short–medium range of distances (i.e., $1 < r^* < 2.5$) the equilateral functions adopt shapes simpler than the pair radial functions. This effect turns out to be much more remarkable for cI16 state points, which show quite a convoluted peak/valley behavior. Accordingly, for the purposes of monitoring the onset of crystallization and/or characterizing irregular solid phases in general, triplet centroid information may advantageously complement the usual pair level information.

  (c) PIMC calculations of solid centroid triplet structures converge slowly with increasing quantum effects, which contrasts with the more rapid convergence of the centroid pair calculations. This fact should be kept in mind when studying centroid triplets in high-density solid phases at low temperatures.

(5) Finally, one must dwell a little more on the (mechanically stable) QHS–cI16 phase that, as is shown in this work and [67], arises for lower densities than in the classical case. Once the question of its appearance from the PIMC evolution of perfect BCC lattices has been settled, there are no symmetry problems related to the calculations of cI16 free energies [67]. The selection of an appropriate reference system (Einstein crystal [66,67]) can be well defined now [80], and the way to computational studies of stability is open. Although there is every reason for believing that, as in the classical case [80], quantum–cI16 is metastable with respect to FCC (or to HCP = hexagonal close-packed) at low temperatures, due to the cI16 higher energies and pressures [67], the assessment of such behavior seems highly valuable. In this connection, one notes the potential QHS-cI16 relations to (i) the high-pressure solid–solid transitions in alkali metals at low temperatures and (ii) the special responses to external fields that these solid structures, which are less tight than FCC (or HCP), might exhibit.

There is work in progress to tackle the issues raised above and to identify some essential facts associated with quantum condensed matter triplets in the real and the Fourier spaces.

Supplementary Materials: The following are available online at http://www.mdpi.com/1099-4300/22/12/1338/s1. SupMat2_Entropy.zip. File1: LMS_SupMat_20S_X1.pdf, contents: PIMC-$g_2(r)$ for bcc–qIII and cI16, PIMC convergence, Structure factors values for perfect cI16 lattices, PIMC salient features on the fluid–FCC coexistence line, complete set of fluid pair radial functions (Figure), cI16 and FCC isosceles correlations, comparison between FCC and cI16 equilateral results (Figure). Triplet functions at ($\rho_N^* = 0.672$, $\lambda_B^* = 0.4$): File2: LMS_SupMat_20S_zgcm3_l4.r672 (centroids), and File3: LMS_SupMat_20S_zget3_l4.r672 (instantaneous).

Funding: This research received no external funding.

Conflicts of Interest: The author declares no conflict of interest.

References

1. Kirkwood, J.G. Statistical Mechanics of Fluid Mixtures. *J. Chem. Phys.* **1935**, *3*, 300–313. [CrossRef]
2. Abe, R. On the Kirkwood Superposition Approximation. *Prog. Theor. Phys.* **1959**, *21*, 421–430. [CrossRef]

3. Jackson, H.W.; Feenberg, E. Perturbation Method for Low States of a Many-Particle Boson System. *Ann. Phys.* **1961**, *15*, 266–295. [CrossRef]
4. Jackson, H.W.; Feenberg, E. Energy Spectrum of Elementary Excitations in Helium-II. *Rev. Mod. Phys.* **1962**, *34*, 686–693. [CrossRef]
5. Egelstaff, P.A.; Page, D.I.; Heard, C.R.T. Experimental Study of the Triplet Correlation Function for Simple Liquids. *Phys. Lett.* **1969**, *30A*, 376–377. [CrossRef]
6. Raveché, H.J.; Mountain, R.D. Three-Body Correlations in Simple Dense Fluids. *J. Chem. Phys.* **1970**, *53*, 3101–3107. [CrossRef]
7. Raveché, H.J.; Mountain, R.D. Three Atom Correlations in Liquid Neon. *J. Chem. Phys.* **1972**, *57*, 3987–3992. [CrossRef]
8. Raveché, H.J.; Mountain, R.D. Structure Studies in Liquid ^{4}He. *Phys. Rev. A* **1974**, *9*, 435–447. [CrossRef]
9. Gubbins, K.E.; Gray, C.G.; Egelstaff, P.A. Thermodynamic Derivatives of Correlation Functions. *Mol. Phys.* **1978**, *35*, 315–328. [CrossRef]
10. Haymet, A.D.J.; Rice, S.A.; Madden, W.G. An Accurate Integral Equation for the Pair and Triplet Distribution Functions of a Simple Liquid. *J. Chem. Phys.* **1981**, *74*, 3033–3041. [CrossRef]
11. Montfrooij, W.; de Graaf, L.A.; van den Bosch, P.J.; Soper, A.K.; Howells, W.S. Density and Temperature Dependence of the Structure Factor of Dense Fluid Helium. *J. Phys. Condens. Matter* **1991**, *3*, 4089–4096. [CrossRef]
12. Whitlock, P.A.; Chester, G.V. Three-Body Correlations in Liquid and Solid ^{4}He. *Phys. Rev. B* **1987**, *35*, 4719–4727. [CrossRef] [PubMed]
13. Feynman, R.P. *Statistical Mechanics*; Benjamin/Cummings: Reading, MA, USA, 1972; ISBN 0805325093.
14. Sesé, L.M. Triplet Correlations in the Quantum Hard-Sphere Fluid. *J. Chem. Phys.* **2005**, *123*, 104507. [CrossRef]
15. Sesé, L.M. Computational Study of the Structures of Gaseous Helium-3 at Low Temperature. *J. Phys. Chem. B* **2008**, *112*, 10241–10254. [CrossRef] [PubMed]
16. Sesé, L.M. A Study of the Pair and Triplet Structures of the Quantum Hard-Sphere Yukawa Fluid. *J. Chem. Phys.* **2009**, *130*, 074504. [CrossRef]
17. Sesé, L.M. On Static Triplet Structures in Fluids with Quantum Behavior. *J. Chem. Phys.* **2018**, *148*, 102312. [CrossRef]
18. Sesé, L.M. Computation of Static Quantum Triplet Structure Factors of Liquid Para-Hydrogen. *J. Chem. Phys.* **2018**, *149*, 124507. [CrossRef]
19. Cencek, W.; Patkowski, K.; Szalewicz, K. Full-Configuration-Interaction Calculation of Three-Body Nonadditive Contribution to Helium Interaction Potential. *J. Chem. Phys.* **2009**, *131*, 064105. [CrossRef]
20. Nettelton, R.E.; Green, M.S. Expression in Terms of Molecular Distribution Functions for the Entropy Density in an Infinite System. *J. Chem. Phys.* **1958**, *29*, 1365–1370. [CrossRef]
21. Raveché, H.J. Entropy and Molecular Correlation Functions in Open Systems. I. Derivation. *J. Chem. Phys.* **1971**, *55*, 2242–2250. [CrossRef]
22. Giaquinta, P.V.; Giunta, G. About Entropy and Correlations in a Fluid of Hard Spheres. *Physica A* **1992**, *187*, 145–158. [CrossRef]
23. Ferziger, J.H.; Leonard, A. Multiple Scattering of Neutrons in the Static Approximation. *Phys. Rev.* **1962**, *128*, 2188–2190. [CrossRef]
24. Barrat, J.L.; Hansen, J.P.; Pastore, G. On the Equilibrium Structure of Dense Fluids Triplet. Correlations, Integral Equations and Freezing. *Mol. Phys.* **1988**, *63*, 747–767. [CrossRef]
25. Evans, R. Density Functionals in the Theory of Nonuniform Fluids. In *Fundamentals of Inhomogeneous Fluids*; Henderson, D., Ed.; Marcel Dekker: New York, NY, USA, 1992; pp. 85–175, ISBN 978-0824787110.
26. Götze, W. *Complex Dynamics of Glass-Forming Liquids*; Oxford University Press: Oxford, UK, 2009, ISBN 9780199656141.
27. Sciortino, F.; Kob, W. Debye-Waller Factor of Liquid Silica: Theory and Simulation. *Phys. Rev. Lett.* **2001**, *86*, 648–651. [CrossRef] [PubMed]
28. Markland, T.E.; Morrone, J.A.; Miyazaki, K.; Berne, B.J.; Reichman, D.R.; Rabani, E. Theory and Simulation of Quantum Glass Forming Liquids. *J. Chem. Phys.* **2012**, *136*, 074511. [CrossRef] [PubMed]
29. Egelstaff, P.A. The Structure of Simple Liquids. *Annu. Rev. Phys. Chem.* **1973**, *24*, 159–187. [CrossRef]

30. Ploetz, E.A.; Smith, P.E. Particle and Energy Pair and Triplet Correlations in Liquids and Liquid Mixtures from Experiment and Simulation. *J. Phys. Chem. B* **2015**, *119*, 7761–7777. [CrossRef]
31. Chandler, D.; Wolynes, P.G. Exploiting the Isomorphism Between Quantum Theory and Classical Statistical Mechanics of Polyatomic Fluids. *J. Chem. Phys.* **1981**, *74*, 4078–4095. [CrossRef]
32. Berne, B.J.; Thirumalai, D. On the Simulation of Quantum Systems: Path Integral Methods. *Annu. Rev. Phys. Chem.* **1986**, *37*, 401–424. [CrossRef]
33. Ceperley, D.M. Path Integrals in the Theory of Condensed Helium. *Rev. Mod. Phys.* **1995**, *67*, 279–355. [CrossRef]
34. Suzuki, M. New Scheme of Hybrid Exponential Product Formulas with Applications to Quantum Monte-Carlo Simulations. In *Computer Simulation Studies in Condensed Matter Physics VIII*; Landau, D.P., Mon, K.K., Schüttler, H.B., Eds.; Springer-Verlag: Berlin, Germany, 1995; pp. 169–174. ISBN 978-3-642-79991-4.
35. Cao, J.; Berne, B.J. A New Quantum Propagator for Hard Sphere and Cavity Systems. *J. Chem. Phys.* **1992**, *97*, 2382–2385. [CrossRef]
36. Grüter, P.; Ceperley, D.; Laloë, F. Critical Temperature of Bose-Einstein Condensation of Hard-Spere Gases. *Phys. Rev. Lett.* **1997**, *79*, 3549–3552. [CrossRef]
37. Chin, S.A. A Symplectic Integrators from Composite Operator Factorizations. *Phys. Lett.* **1997**, *226*, 344–348. [CrossRef]
38. Jang, S.; Jang, S.; Voth, G.A. Applications of Higher-Order Composite Factorization Schemes in Imaginary Time Path Integral Simulations. *J. Chem. Phys.* **2001**, *115*, 7832–7842. [CrossRef]
39. Sesé, L.M. The Compressibility Theorem for Quantum Simple Fluids at Equilibrium. *Mol. Phys.* **2003**, *101*, 1455–1468. [CrossRef]
40. Li, Y.; Miller, H. Different Time Slices for Different Degrees of Freedom in Feynman Path Integration. *Mol. Phys.* **2005**, *103*, 203–208. [CrossRef]
41. Boninsegni, M.; Prokof'ev, N.V.; Svistunov, B.V. Worm Algorithm and Diagrammatic Monte Carlo: A New Approach to Continuous-Space Path Integral Monte Carlo Simulations. *Phys. Rev E* **2006**, *74*, 036701. [CrossRef]
42. Boninsegni, M. Quantum Statistics and the Momentum Distribution of Liquid Parahydrogen. *Phys. Rev. B* **2009**, *79*, 174203. [CrossRef]
43. Liberatore, E.; Pierleoni, C.; Ceperley, D.M. Liquid-Solid Transition in Fully Ionized Hydrogen at Ultra-High Pressures. *J. Chem. Phys.* **2011**, *134*, 184505. [CrossRef]
44. Pérez, A.; Tuckerman, M. Improving the Convergence of Closed and Open Path Integral Molecular Dynamics via Higher-Order Trotter Factorization Schemes. *J. Chem. Phys.* **2011**, *135*, 064104. [CrossRef]
45. Sinitskiy, A.V.; Voth, G.A. A Reductionist Perspective on Quantum Statistical Mechanics: Coarse-Graining of Path Integrals. *J. Chem. Phys.* **2015**, *143*, 094104. [CrossRef] [PubMed]
46. Mielke, S.L.; Truhlar, D.G. Improved Methods for Feynman Path Integral Calculations and their Application to Calculate Converged Vibrational-Rotational Partition Functions, Free Energies, Enthalpies, and Heat Capacities for Methane. *J. Chem. Phys.* **2015**, *142*, 044105. [CrossRef] [PubMed]
47. Sesé, L.M. Path Integrals and Effective Potentials in the Study of Monatomic Fluids at Equilibrium. In *Advances in Chemical Physics*; Rice, S.A., Dinner, A.R., Eds.; Wiley: New York, NY, USA, 2016; Volume 160, pp. 49–158. [CrossRef]
48. Sesé, L.M. Path-Integral and Ornstein-Zernike Computations of Quantum Fluid Structures Under Strong Fluctuations. *AIP Adv.* **2017**, *7*, 025204. [CrossRef]
49. Sesé, L.M. Path-Integral and Ornstein-Zernike Study of Quantum Fluid Structures on the Crystallization Line. *J. Chem. Phys.* **2016**, *144*, 094505. [CrossRef]
50. Cendagorta, J.R.; Bacic, Z.; Tuckerman, M. An Open-Chain Imaginary-Time Path-Integral Sampling Approach to the Calculation of Approximate Symmetrized Quantum Time Correlation Functions. *J. Chem. Phys.* **2018**, *148*, 102340. [CrossRef]
51. Han, Y.; Jin, J.; Wagner, J.W.; Voth, G.A. Quantum Theory of Multiscale Coarse-Graining. *J. Chem. Phys.* **2018**, *148*, 102335. [CrossRef]
52. Rillo, G.; Morales, M.A.; Ceperley, D.M.; Pierleoni, C. Coupled Electron-Ion Monte Carlo Simulation of Hydrogen Molecular Crystals. *J. Chem. Phys.* **2018**, *148*, 102314. [CrossRef]
53. Herrero, C.P.; Ramírez, R. Thermal Properties of Graphene from Path-Integral Simulations. *J. Chem. Phys.* **2018**, *148*, 102302. [CrossRef]

54. Boninsegni, M. Kinetic Energy and Momentum Distribution of Isotopic Liquid Helium Mixtures. *J. Chem. Phys.* **2018**, *148*, 102308. [CrossRef]
55. Schran, C.; Uhl, F.; Behler, J.; Marx, D. High-Dimensional Neural Network Potentials for Solvation: The case of Protonated Water Clusters in Helium. *J. Chem. Phys.* **2018**, *148*, 102310. [CrossRef]
56. Allen, M.P.; Tildesley, D.J. *Computer Simulation of Liquids*; Clarendon: Oxford, UK, 1989, ISBN 9780198556459.
57. Tanaka, M.; Fukui, Y. Simulation of the Three-Particle Distribution Function in a Long-Range Oscillatory Potential Liquid. *Prog. Theor. Phys.* **1975**, *53*, 1547–1565. [CrossRef]
58. Baranyai, A.; Evans, D.J. Three-Particle Contribution to the Configurational Entropy of Simple Fluids. *Phys. Rev. A* **1990**, *42*, 849–857. [CrossRef] [PubMed]
59. Bildstein, B.; Kahl, G. Triplet Correlation Functions for Hard-Spheres: Computer Simulation Results. *J. Chem. Phys.* **1994**, *100*, 5882–5893. [CrossRef]
60. Jorge, S.; Lomba, E.; Abascal, J.L.F. Theory and Simulation of the Triplet Structure Factor and Triplet Direct Correlation Functions in Binary Mixtures. *J. Chem. Phys.* **2002**, *116*, 730–736. [CrossRef]
61. Trotter, H.F. Approximation of Semi-Groups of Operators. *Pacific J. Math.* **1958**, *8*, 887–919. [CrossRef]
62. Herman, M.F.; Bruskin, E.J.; Berne, B.J. On Path Integral Monte Carlo Simulations. *J. Chem. Phys.* **1982**, *76*, 5150–5155. [CrossRef]
63. Sesé, L.M. An Application of the Self-Consistent Variational Effective Potential Against the Path-Integral to Compute Equilibrium Properties of Quantum Simple Fluids. *Mol. Phys.* **1999**, *97*, 881–896. [CrossRef]
64. Marx, D.; Müser, M.H. Path Integral Simulations of Rotors: Theory and Applications. *J. Phys. Condens. Matter* **1999**, *11*, R117–R155. [CrossRef]
65. Müser, M.H.; Berne, B.J. Path-Integral Monte Carlo Scheme for Rigid Tops: Application to the Quantum Rotator Phase Transition in Solid Methane. *Phys. Rev. Lett.* **1996**, *77*, 2638–2641. [CrossRef]
66. Runge, K.J.; Chester, G.V. Solid-Fluid Phase Transition of Quantum Hard Spheres at Finite Temperature. *Phys. Rev. B* **1988**, *38*, 135–162. [CrossRef]
67. Sesé, L.M. Path Integral Monte Carlo Study of Quantum-Hard Sphere Solids. *J. Chem. Phys.* **2013**, *139*, 044502. [CrossRef] [PubMed]
68. Vega, C.; Conde, M.M.; McBride, C.; Abascal, J.L.F.; Noya, E.G.; Ramírez, R.; Sesé, L.M. Heat Capacity of Water: A Signature of Nuclear Quantum Effects. *J. Chem. Phys.* **2010**, *132*, 046101. [CrossRef] [PubMed]
69. Herrero, C.; Ramírez, R. Path-Integral Simulation of Solids. *J. Phys. Condens. Matter* **2014**, *26*, 233201. [CrossRef] [PubMed]
70. Ceperley, D.M. Path-Integral Calculations of Normal Liquid ^{3}He. *Phys. Rev. Lett.* **1992**, *69*, 331–334. [CrossRef]
71. Hansen, J.-P.; Levesque, D.; Schiff, D. Fluid-Solid Phase Transition of a Hard-Sphere Bose System. *Phys. Rev. A* **1971**, *3*, 776–780. [CrossRef]
72. Kalos, M.H.; Levesque, D.; Verlet, L. Helium at Zero Temperature with Hard-Sphere and Other Forces. *Phys. Rev. A* **1974**, *9*, 2178–2195. [CrossRef]
73. Dang, L.; Boninsegni, M. Phases of Lattice Hard-Core Bosons in a Periodic Superlattice. *Phys. Rev. B* **2010**, *81*, 224502. [CrossRef]
74. Pusey, P.N.; van Megen, W. Phase Behaviour of Concentrated Suspensions of Nearly Hard Colloidal Spheres. *Nature* **1986**, *320*, 340–342. [CrossRef]
75. Ho, H.M.; Lin, B.; Rice, S.A. Three-Particle Correlation Functions of Quasi-Two-Dimensional One-Component and Binary Colloid Suspensions. *J. Chem. Phys.* **2006**, *125*, 184715. [CrossRef]
76. Hanfland, M.; Syassen, K.; Christensen, N.E.; Novikov, D.L. New High-Pressure Phases of Lithium. *Nature* **2000**, *408*, 174–178. [CrossRef]
77. McMahon, M.I.; Gregoryanz, E.; Lundegaard, L.F.; Loa, I.; Guillaume, C.; Nelmes, R.J.; Kleppe, A.K.; Amboage, M.; Wilhelm, H.; Jephcoat, A.P. Structure of Sodium Above 100 GPa by Single-Crystal X-Ray Diffraction. *Proc. Nat. Acad. Sci. USA* **2007**, *44*, 17297–17299. [CrossRef] [PubMed]
78. Sesé, L.M. Path-Integral Monte Carlo Study of the Structural and Mechanical Properties of Quantum fcc and bcc Hard-Sphere Solids. *J. Chem. Phys.* **2001**, *114*, 1732–1744. [CrossRef]
79. Curtin, W.A.; Runge, K. Weighted-Density-Functional and Simulation Studies of the bcc Hard-Sphere Solid. *Phys. Rev. A* **1987**, *35*, 4755–4762. [CrossRef] [PubMed]
80. Warshavsky, W.B.; Ford, D.M.; Monson, P.A. On the Mechanical Stability of the Body-Centered Cubic Phase and the Emergence of a Metastable cI16 Phase in Classical Hard-Sphere Solids. *J. Chem. Phys.* **2018**, *148*, 024502. [CrossRef]

81. Cao, J.; Voth, G.A. The Formulation of Quantum Statistical Mechanics Based on the Feynman Path Centroid Density. I. Equilibrium Properties. *J. Chem. Phys.* **1994**, *100*, 5093–5105. [CrossRef]
82. Cao, J.; Voth, G.A. Semiclassical Approximations to Quantum Dynamical Time Correlation Functions. *J. Chem. Phys.* **1996**, *104*, 273–285. [CrossRef]
83. Miura, S.; Okazaki, S.; Kinugawa, K. A Path Integral Centroid Molecular Dynamics Study of Nonsuperfluid Liquid Helium-4. *J. Chem. Phys.* **1999**, *110*, 4523–4532. [CrossRef]
84. Ramírez, R.; López-Ciudad, T.; Noya, J.C. Feynman Effective Classical Potential in the Schrödinger Formulation. *Phys. Rev. Lett.* **1998**, *81*, 3303–3306. [CrossRef]
85. Ramírez, R.; López-Ciudad, T. The Schrödinger formulation of the Feynman path centroid density. *J. Chem. Phys.* **1999**, *111*, 3339–3348. [CrossRef]
86. Sesé, L.M. On the Accurate Direct Computation of the Isothermal Compressibility for Normal Quantum Simple Fluids: Application to Quantum Hard Spheres. *J. Chem. Phys.* **2012**, *136*, 244504. [CrossRef]
87. Hemmer, P.C. The Hard-Core Quantum Gas at High Temperature. *Phys. Lett. A* **1968**, *27*, 377–378. [CrossRef]
88. Jancovici, B. Quantum-Mechanical Equation of State of a Hard-Sphere Gas at High Temperature. II*. *Phys. Rev.* **1969**, *184*, 119–123. [CrossRef]
89. Gibson, W.G. Quantum Corrections to the Properties of a Dense Fluid with Non-Analytic Intermolecular Potential Function. II Hard Spheres. *Mol. Phys.* **1975**, *30*, 13–30. [CrossRef]
90. Yoon, B.J.; Scheraga, H.A. Monte Carlo Simulation of the Hard-Sphere Fluid with Quantum Correction and Estimate of its Free Energy. *J. Chem. Phys.* **1988**, *88*, 3923–3933. [CrossRef]
91. Boninsegni, M.; Pierleoni, C.; Ceperley, D.M. Isotopic Shift of Helium Melting Pressure: Path Integral Monte Carlo Study. *Phys. Rev. Lett.* **1994**, *72*, 1854–1857. [CrossRef]
92. Moroni, S.; Pederiva, F.; Fantoni, S.; Boninsegni, M. Equation of State of Solid ^{3}He. *Phys. Rev. Lett.* **2000**, *84*, 2650–2653. [CrossRef]
93. Barnes, A.L.; Hinde, R.J. Three-Body Interactions and the Elastic Constants of hcp Solid ^{4}He. *J. Chem. Phys.* **2017**, *147*, 114504. [CrossRef]
94. Baxter, R.J. Ornstein-Zernike Relation for a Disordered Fluid. *Aust. J. Phys.* **1968**, *21*, 563–569. [CrossRef]
95. Dixon, M.; Hutchinson, J.P. A Method for the Extrapolation of Pair Distribution Functions. *Mol. Phys.* **1977**, *33*, 1663–1670. [CrossRef]
96. Baumketner, A.; Hiwatari, Y. Finite-Size Dependence of the Bridge Function Extracted from Molecular Dynamics Simulations. *Phys. Rev. E* **2001**, *63*, 061201. [CrossRef]
97. Sesé, L.M. Determination of the Quantum Static Structure Factor of Liquid Neon within the Feynman-Hibbs Picture. *Mol. Phys.* **1996**, *89*, 1783–1802. [CrossRef]
98. Sesé, L.M. Thermodynamic and Structural Properties of the Path-Integral Quantum Hard-Sphere Fluid. *J. Chem. Phys.* **1998**, *108*, 9086–9097. [CrossRef]
99. Melrose, J.R.; Singer, K. An Investigation of Supercooled Lennard-Jones Argon by Quantum Mechanical and Classical Monte Carlo Simulation. *Mol. Phys.* **1989**, *66*, 1203–1214. [CrossRef]
100. Mandell, M.J.; McTague, J.P.; Rahman, A. Crystal Nucleation in a Three-Dimensional Lennard-Jones System. II. Nucleation Kinetics for 256 and 500 Particles. *J. Chem. Phys.* **1977**, *66*, 3070–3075. [CrossRef]
101. Steinhardt, P.J.; Nelson, D.R.; Ronchetti, M. Bond-Orientational Order in Liquids and Glasses. *Phys. Rev. B* **1983**, *28*, 784–805. [CrossRef]

Publisher's Note: MDPI stays neutral with regard to jurisdictional claims in published maps and institutional affiliations.

© 2020 by the author. Licensee MDPI, Basel, Switzerland. This article is an open access article distributed under the terms and conditions of the Creative Commons Attribution (CC BY) license (http://creativecommons.org/licenses/by/4.0/).

Article

The Molecular Theory of Liquid Nanodroplets Energetics in Aerosols

Sergii D. Kaim

Faculty of Electrical Engineering Automatic Control and Informatics, Opole University of Technology, ul. Prószkowska 76, 45-758 Opole, Poland; s.kaim@po.edu.pl; Tel.: +48-77-537-809-228

Abstract: Studies of the coronavirus SARS-CoV-2 spread mechanisms indicate that the main mechanism is associated with the spread in the atmosphere of micro- and nanodroplets of liquid with an active agent. However, the molecular theory of aerosols of microdroplets in gases remains poorly developed. In this work, the energy properties of aerosol nanodroplets of simple liquids suspended in a gas were studied within the framework of molecular theory. The three components of the effective aerosol Hamiltonian were investigated: (1) the interaction energy of an individual atom with a liquid nanodroplet; (2) the surface energy of liquid nanodroplet; and (3) the interaction energy of two liquid nanodroplets. The size dependence of all contributions was investigated. The pairwise interparticle interactions and pairwise interparticle correlations were accounted for to study the nanodroplet properties using the Fowler approximation. In this paper, the problem of the adhesion energy calculation of a molecular complex and a liquid nanodroplet is discussed. The derived effective Hamiltonian is generic and can be used for the cases of multicomponent nano-aerosols and to account for particle size distributions.

Keywords: nanodroplet aerosols; the effective Hamiltonian; surface energy; atom–nanodroplet interaction energy; interaction energy of two nanodroplets; size dependence; adhesion energy of a molecular complex and a liquid nanodroplet

Citation: Kaim, S.D. The Molecular Theory of Liquid Nanodroplets Energetics in Aerosols. *Entropy* **2021**, *23*, 13. https://dx.doi.org/10.3390/e23010013

Received: 31 October 2020
Accepted: 21 December 2020
Published: 24 December 2020

Publisher's Note: MDPI stays neutral with regard to jurisdictional claims in published maps and institutional affiliations.

Copyright: © 2020 by the author. Licensee MDPI, Basel, Switzerland. This article is an open access article distributed under the terms and conditions of the Creative Commons Attribution (CC BY) license (https://creativecommons.org/licenses/by/4.0/).

1. Introduction

The rapid spread of coronavirus SARS-CoV-2 has become an investigation subject for numerous scientists. The existing data exposed the ability of a virus to be transmitted in an airborne manner as dispersed droplets that contain the infective agent [1–6]. Airborne transmission is defined by the World Health Organization (WHO) as the spread of invective agents through suspended droplets in the air, which stay infective for long periods of time and may travel long distances [7].

An important physical aspect in the problem of virus spread is the interaction of nanoparticles (virions) and nanodroplets with molecular structures of different media. For these types of problems, an important characteristic is the size dependence of energetic properties of nanodroplets and nanoparticles at interactions with different media structures. The calculations of the adhesion energy of nanoparticles to the different structures, in addition to the calculations of energetic characteristics for aerosol nanodroplets, require an application of statistical physics methods. An investigation of equilibrium and nonequilibrium properties of droplets and aerosols with liquid nanodroplets can be performed within the framework of classical statistical mechanics. The nanodroplets may reveal implicit collective properties and self-organization into structures at a macroscopic level [8]. The behavior of an isolated nanodroplet can be simulated by means of molecular dynamics [9].

The typical complications in the theoretical analysis of nanosystems are conditioned by the necessity to account for the surface terms for all of its equilibrium and nonequilibrium properties. At nanometer scales, an abrupt change of all system characteristics takes place near the surface. The intermolecular forces act at similar scales.

The contemporary state of theoretical studies of the structure and properties of liquid nanodroplets within the framework of molecular–kinetic theory has been summarized in the works [10–12]. A statistical approach to the study of volumetric properties of equilibrium and nonequilibrium homogeneous systems shows the importance of accounting for the paired interparticle interactions and correlations, which contain determinative terms in all properties [13–16]. For inhomogeneous liquids in statistical theory, it is important to take into account the one-particle distribution functions and the effective one-particle potentials, which are determined by paired interparticle interactions and correlations. The contribution of the one-particle effects to the properties of inhomogeneous systems corresponds to the accounting of surface terms.

The current study, using the correlation theory of inhomogeneous liquids, investigated the equilibrium properties of nanodroplets as components of aerosol systems. Thus, there was a need to use approximations that allowed a reasonable comparison of theoretical and experimental results to be made. The properties of aerosol systems were determined by their interacting components, i.e., nanodroplets and gas. There exist many phenomena at the molecular level that lack explanation. Under these conditions, it is difficult to predict the behavior of nano-objects and their assemblies, or to control these nanosystems. For the investigated aerosol systems, a fundamental role is played by the energy characteristics of the separate nanodroplets, their interaction energies with the surrounding gas, and the paired interactions of nanodroplets. Investigation of the pointed energy characteristics is a major focus of the current article.

2. The Calculation of the Molecule Interaction Energy with Liquid Nanodroplet

Formulation and resolution of the problems related to the interaction of isolated atoms and molecules with the heterogeneity of the condensed system (surface, new phase origins, phase transition fronts) are essential in constructing a microscopic theory of first-order phase transitions, in addition to a microscopic theory of equilibrium and kinetic properties of the surface and interphase boundaries. A review of the current state of the problems of atom interactions with an inhomogeneous environment can be found in [10–12,17]. However, the problems of size dependencies of the interaction energy of atoms and nanosized condensed systems, with accounting for interatomic correlations, remain unsolved.

In statistical mechanics of condensed matter, as a rule, the potential interaction energy of two atoms or two molecules is assumed to be known. An approximation of pair additive potentials is widely used to describe the energies of interatomic interactions [10–16]. Among the most used potentials of interatomic interaction, Lennard–Jones potential and its generalizations, Morse potential, hard-sphere potential, and soft-sphere potential should be noted [10–16].

Calculation of the interaction energy of macroscopic bodies with different geometries in continual approximation, including accounting for van der Waals forces, is described in [11,12]. The calculations of atom interaction energy with an object outside the framework of continual approximation, when it is necessary to account for the repulsion of atoms and pair correlation in its position, have attracted significant interest. For an interaction of isolated atoms with solid objects, the continual approximation is correct and accounts for the repulsion of an atom from the solid body atoms (without accounting for an atomistic structure of the solid body and the possibility of atoms penetrating into the body). However, in the case of atom–liquid interaction, this approach is insufficient. In the equilibrium system of liquid–vapor, an interchange of atoms between both phases takes place, and the equilibrium is dynamic. To take account of an interchange between two phases, it is necessary to equally account for a discrete structure of both the vapor and liquid. A mathematical technique to describe this process should be similar for both phases.

A wide overview of the literature regarding the interaction of atoms and macroscopic bodies [11] points to a range of unsolved problems, which are essential for understanding the processes in the nanoscale systems. A topical problem is the calculation of the interaction energy of atoms with droplets and the solid bodies of nanoscale size, including accounting for the surface curvature and repulsive effects, and corresponding correlations in their positions.

The goal of the current section is to present the model calculations of the interaction energy between the atoms and nanodroplets of a simple liquid. The foundation of these calculations is based on the expression for the energy of an inhomogeneous liquid within the framework of the distribution function method of groups of particles, which takes into account all paired interatomic interactions and correlations. The paired interatomic interactions were described using the Lennard–Jones potential. The structure of the droplets was modeled using a Fowler approximation (the step profile of atoms' density in the droplets) [18,19]. The calculations are performed for the cases in which the atom is located outside, inside, and on the surface of the droplet. For the geometry dispositions of the atom and droplet, which require accounting for the paired interatomic correlations, pair distribution functions within the framework of thermodynamic perturbation theory were used.

The potential energy of an inhomogeneous system of a pair of interacting particles, located in a volume V, can be written as [10,19]:

$$E = \frac{1}{2} n_0^2 \int_V d^3 r_1 \int_V d^3 r_2 F_2(\mathbf{r}_1, \mathbf{r}_2) \Phi(\mathbf{r}_1, \mathbf{r}_2) \tag{1}$$

where $\Phi(\mathbf{r}_1, \mathbf{r}_2)$ is an interaction energy of the pair of atoms; is a pair distribution function of atoms inside an inhomogeneous liquid in a canonical Gibbs ensemble; and n_0 is the density of the number of particles. Formula (1) takes into account all paired interparticle interactions and correlations for the energy of the inhomogeneous liquid. The interaction energy of the volume element dV of the liquid with the remainder of the liquid can be written as:

$$E_{el-liq} = n_0^2 dV_1 \int_V d^3 r_2 F_2(\mathbf{r}_1, \mathbf{r}_2) \Phi(\mathbf{r}_1, \mathbf{r}_2) \tag{2}$$

By choosing the volume element from the condition $dV_1 n_0 = 1$, we obtain an expression for the interaction energy of an isolated atom with an inhomogeneous liquid. For the model calculations, a central symmetry potential of paired interaction of the atoms was used, and the following approximation for the pair distribution function inside the droplet:

$$F_2(\mathbf{r}_1, \mathbf{r}_2) \cong F_2^{(0)}(|\mathbf{r}_1 - \mathbf{r}_2|) \Theta(a - r_1) \Theta(a - r_2), \tag{3}$$

where $F_2^{(0)}(|\mathbf{r}_1 - \mathbf{r}_2|)$ is a pair distribution function of atoms in homogeneous liquid; $\Theta(x)$ is the Heaviside step-function; a is the radius of the droplet.

Figure 1 shows a droplet and atom at the point A. The radius vector $\mathbf{R}_1$ indicates the location of the atom, interacting with the droplet, and radius vector $\mathbf{R}_2$ indicates the location of the volume element. By introducing new integration variables via relation $\mathbf{R}_2 - \mathbf{R}_1 \equiv \mathbf{R}_{12}$ and using spherical coordinates for integration (Figure 1), we obtain the following expression for the atom–droplet interaction:

$$E_{a-d}(R_1, a) = n_0 \int_0^\infty dR R^2 \int_0^\pi \sin\Theta d\Theta \int_0^{2\pi} d\varphi \Phi(R) F_2^{(0)}(R) \times \\ \times \Theta\left(a - (R_1^2 + R^2 + 2R_1 R \cos\Theta)^{1/2}\right). \tag{4}$$

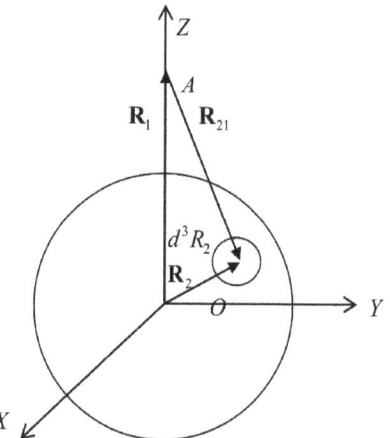

Figure 1. Geometry of atom and droplet location.

After integration over spherical variables, the expression for the atom–droplet interaction energy takes the form:

$$E_{a-d}(R_1,a) = \frac{\pi n_0}{R_1}\Theta(R_1-a)\int_{R-a}^{R+a} dR R\Phi(R)F_2^{(0)}(R) \times \\ \times [a^2 - R_1^2 - R^2 + 2RR_1] \quad (5)$$

where R_1 is the distance from the atom to the center of the droplet. The obtained expression is valid for the distances $R_1 \geq a$. From (5), we can see that the atom–droplet interaction potential is dependent on the geometrical size of the droplet and the thermodynamic parameters, such as density of the number of atoms and temperature. Formally, $E_{a-d}(R_1,a)$ is a function of the potential of the paired interparticle interaction and pair distribution function of the atoms. The derived Expression (5) takes into account the kinematic conditions of the atoms' arrangement outside the droplet.

If the atom is located at the surface of the droplet, then $R_1 = a$ and the interaction energy expression takes the form:

$$E_{a-d}(a,a) = \frac{\pi n_0}{a}\int_0^{2a} dR R F_2^{(0)}(R)\Phi(R)(2aR - R^2). \quad (6)$$

In this case, from (6), it is clearly seen that the atom–droplet interaction energy takes finite values, unlike in the case of continual approximations, in which no body structure is taken into account and density is assumed to be constant [11,12]. In the continual approximations, the atom energy at the surface of the droplet tends to infinity due to the repulsive forces acting on the atom from the droplet.

For the case in which the atom is located inside the droplet, similar calculations allow the interaction energy of the atom and droplet to be obtained:

$$E_{a-d}(R_1,a) = 4\pi n_0 \Theta(a-R_1)\int_0^{a-R_1} dR R^2 \Phi(R) F_2^{(0)}(R) + \\ + \frac{\pi n_0}{R_1}\Theta(a-R_1)\int_{a-R_1}^{a+R_1} dR R\Phi(R)F_2^{(0)}(R)\left[a^2 - (R-R_1)^2\right] \quad (7)$$

If we choose in Expression (7), then we obtain Expression (6).

If the atom is located at the center of the droplet, then its interaction energy with the droplet will have the form:

$$E_{a-d}(0,a) = 4\pi n_0 \int_0^a dR R^2 \Phi(R) F_2^{(0)}(R). \tag{8}$$

In a boundary case, when the droplet radius approaches infinity, from (8), the expression for the atom interaction energy with unbounded liquid can be obtained:

$$E_a = 4\pi n_0 \int_0^\infty dR R^2 \Phi(R) F_2^{(0)}(R). \tag{9}$$

Expressions (8) and (9) have an explicit geometrical meaning.

The expression for the interaction energy of the atom with a semi-bounded liquid in the case of arbitrary distances from the atom to the surface, when it is important that the pair atom–atom correlations in the semi-bounded liquid are accounted for, can be derived from Expression (5) by means of transition $a \to \infty, R_1 - a = d = const$:

$$E_{a-f}(d) = 2\pi n_0 \int_d^\infty dR F_2^{(0)}(R) \Phi(R) R(R-d) \tag{10}$$

where d is a distance from the atom to the surface.

The interaction energy of an atom that is located on a flat surface of a semi-bounded liquid in the case of the Fowler approximation will take a form of a particular case of (10):

$$E_{a-f}(0) = 2\pi n_0 \int_0^\infty dR R^2 F_2^{(0)}(R) \Phi(R). \tag{11}$$

In contrast to the results of continual approximation, the interaction energy (11) is limited. In Expressions (4)–(11) for the interaction energy, the divergence of the corresponding integrals in the accounting for the paired interatomic correlations is absent.

Let us consider a boundary case of atom interaction energy with a nanodroplet of liquid in a continual approximation. In the case of distances $R_1 - a \gg \sigma$, where σ is a characteristic length for the pair distribution function, we can assume $F_2^{(0)} \cong 1$ and integrate using the explicit expression for the atom's paired interaction potential. Thus, in the case of Lennard–Jones potential:

$$\Phi(R) = 4\varepsilon\left[\left(\frac{\sigma}{R}\right)^{12} - \left(\frac{\sigma}{R}\right)^6\right] \tag{12}$$

with the parameters σ and ε, we obtain:

$$\begin{aligned}E_{a-d}(a,R_1) = \frac{4\pi n_0 \varepsilon \sigma^6}{R_1}\Bigg\{&\frac{(R_1^2-a^2)\sigma^6}{10}\left[\frac{1}{(R_1+a)^{10}} - \frac{1}{(R_1-a)^{10}}\right] + \\&+\frac{1}{8}\sigma^6\left[\frac{1}{(R_1+a)^8} - \frac{1}{(R_1-a)^8}\right] - \frac{2}{9}R_1\sigma^6\left[\frac{1}{(R_1+a)^9} - \frac{1}{(R_1-a)^9}\right] + \\&+\frac{(a^2-R_1^2)}{4}\left[\frac{1}{(R_1+a)^4} - \frac{1}{(R_1-a)^4}\right] - \frac{1}{2}\left[\frac{1}{(R_1+a)^2} - \frac{1}{(R_1-a)^2}\right] + \\&+\frac{2}{3}R_1\left[\frac{1}{(R_1+a)^3} - \frac{1}{(R_1-a)^3}\right]\Bigg\}.\end{aligned} \tag{13}$$

The interaction energy of an atom with the semi-infinite liquid in the boundary case $a \to \infty$ can be written as:

$$E_{a-f}(d) = \frac{4\pi}{9} n_0 \sigma^3 \varepsilon \left\{ \frac{1}{5} \left(\frac{\sigma}{d}\right)^9 - \frac{3}{2} \left(\frac{\sigma}{d}\right)^3 \right\}, \qquad (14)$$

where d is the distance between the atom and a flat surface. In the general case, the potential interaction parameters of an atom with a droplet and an atom with semi-infinite liquid are functions of density and temperature of the liquid.

From Expression (14), we can obtain the position of the first zero d_0 of the potential and the value of the potential minimum $\backslash d_{\min}$:

$$d_0 = \frac{\sigma}{\sqrt[6]{3}}, d_{\min} = \sigma. \qquad (15)$$

The depth of the potential well in which the atom near the flat surface moves will take the form:

$$U = -\frac{8\pi n_0 \sigma^3}{9} \varepsilon. \qquad (16)$$

The interaction energy of an atom with a semi-infinite liquid at long distances from the surface is the inverse proportional cube of the distance to the surface and has the following asymptote:

$$E_{a-f} \cong -\frac{\pi n_0 C}{6} \frac{1}{d^3}, \qquad (17)$$

where $C = 4\varepsilon\sigma^6$ is a constant in the Van der Waals potential. Asymptotic behavior:

$$E_{a-f} \sim \frac{1}{d^3} \qquad (18)$$

is obtained within the framework of macroscopic Van der Waals interaction theory (without accounting for delay effects) [20].

The numerical calculations were performed for the normal ^4He at a temperature $T = 2.2K$ and a density $\rho = 147$ kg/m^3, which corresponds to the density of the number of particles $n = 22.1266$ nm^{-3}. Parameters of the Lennard–Jones potential were chosen to be $\sigma = 2.576$Å, $\varepsilon = 10.22K$ [13]. The pair distribution function was modeled by means of the distribution function obtained within the framework of the Barker–Henderson thermodynamic perturbation theory [10,16].

Figure 2 shows the results of calculations in continual approximation using Formula (13) for the interaction energy of an atom with a droplet of liquid helium for different values of droplet radius (curves 1–3) and a semi-bounded liquid (curve 4). The interaction energy of the pair of atoms, which is described by the Lennard–Jones potential (curve 5), is also shown. The calculations show that the position of the minimum of the interaction energy of an atom with droplets and semi-bounded liquids is significantly shifted to the shorter distances in comparison to the interatomic potential $\Phi(R)$. The depth of the potential well increases with the growth of the droplet radius and reaches saturation for the flat interface of the liquid. At a given density of the liquid, the depth of the potential well, even in the case of atom interaction with the flat interface, is less than in the case of atom–atom interaction. In the case of continual approximation, the interaction energy of the atom with its surroundings approaches infinity when an atom approaches a surface, i.e., the atom cannot reach the surface of the liquid. This result is not satisfactory for a liquid; however, to a certain level, it is acceptable for the modeling of atom interaction with a solid body, and it is used in absorption problems. The main disadvantage of the continuum approximation is complete neglect of the discrete surrounding structure effects and correlations between separate atoms with the surrounding atoms.

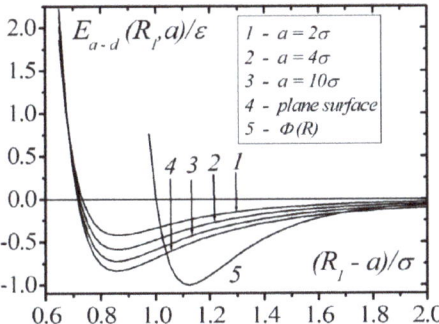

Figure 2. The results of model calculations of the energy of the interaction of atoms and nanodroplets of ^4He with different radiuses.

The calculation results of the interaction energy of an atom with a droplet in the framework of correlation theory in the case when the atom is located on the surface of the droplet are shown in Figure 3. From Figure 3, we can see that at droplet radius vales around $[20 - 30\sigma]$, the energy $E_{a-d}(a,a)$, as a function of the radius, approaches the asymptotic value. In contrast to the continuum model, accounting for the interatomic correlations of the atom on the surface of the droplet demonstrates finite negative values of the energy, which corresponds to the attraction of the atom to the droplet.

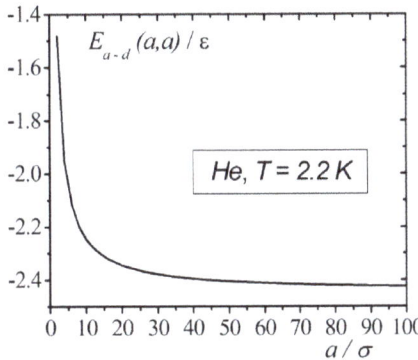

Figure 3. Results of calculations of the interaction energy of an atom with a droplet of ^4He when the atom is located on the surface of the droplet.

Figure 4 shows the calculation results of the interaction energy of an atom with a droplet as a function of the droplet radius $E_{a-d}(0,a)$, for the case when atom is located in the center of the droplet. From the graph in Figure 4, it is seen that at radius values of the order 10σ, the interaction energy of the atom with droplet $E_{a-d}(0,a)$ reaches its asymptotic values for the unbounded liquid; however, it remains finite.

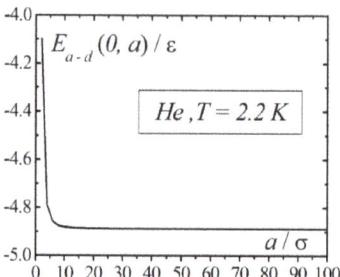

Figure 4. The interaction energy of the atom with a droplet as a function of the droplet radius for the case when the atom is located in the center of the droplet.

Figure 5 shows the calculation results of the atom interaction energy with a liquid helium droplet, including accounting for the correlation effects at variable radius values and arbitrary distances from the atom to the center of the droplet. As can be seen from the graph, the interaction energy of an atom with a droplet exhibits a saturation effect. This effect can be seen in the following circumstances. (1) The droplet radius values are of the order of 4σ, and the number of atoms in the droplet is around 101. A one-atom potential is formed, which covers most of the droplet and corresponds to the asymptotic value for the unbounded liquid. (2) The thickness of the near-surface layer, where the interaction energy varies from its value inside the droplet to the asymptotic value outside the droplet, quickly reaches the values of the order 6σ at radius growth. These results indicate that a majority of atoms inside the droplet with a radius $a > 10\sigma$ are under the influence of self-consistent potential, which is similar to that of the homogeneous unbounded liquid, and the gradient of this potential is located at the near-surface layer of the thickness 6σ. As a result, at radius growth, the thickness of the atom density profile of the near-surface layer quickly reaches values corresponding to the flat surface.

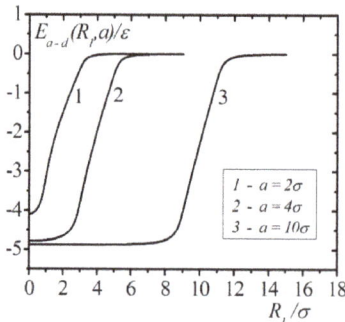

Figure 5. The interaction energy of the atom with the droplet of liquid ^4He, accounting for correlation effects with variable droplet radius values.

For comparison, Figure 6 shows the dependences of the interaction energies of an atom and ^4He nanodroplet of radius 2σ in the continual model and taking into account the correlation effects. The two curves practically coincide only for large distances $R > 2.8\sigma$. At smaller distances, there is a significant difference and in the boundary case $R \to 2\sigma + 0$ in the continual model $E_{a-d}(R, a) \to \infty$.

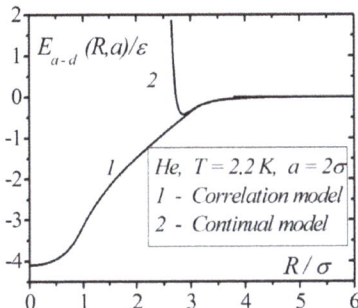

Figure 6. Comparison of the energies of interaction of an atom with a droplet of ^4He with radius $a = 2\sigma$ in the continual model and taking into account correlation effects.

3. Microscopic Theory of Nanodroplet Surface Energy in Fowler Approximation

Let us consider a simple liquid in a volume V, which is described by the Hamiltonian:

$$H = \sum_{i=1}^{N} \frac{\mathbf{P}_i^2}{2M} + \frac{1}{2}\sum_{i\neq j=1}^{N} \Phi(|\mathbf{R}_i - \mathbf{R}_j|), \tag{19}$$

where $\mathbf{P}_i$, M are the impulse and mass of an atom, respectively; $\Phi(|\mathbf{R}_i - \mathbf{R}_j|)$ is the central–symmetric potential of interatomic interaction; and N is the number of atoms. After an averaging within the framework of the distribution function method of groups of particles [10,13–16,19], for the energy of liquid, we obtain:

$$E = \langle H \rangle = E_k + E_p = \tfrac{3}{2} N k_B T +$$
$$+ \frac{N(N-1)}{2V^2} \int_V d^3 R_1 \int_V d^3 R_2 F_2(|\mathbf{R}_1 - \mathbf{R}_2|)\Phi(|\mathbf{R}_1 - \mathbf{R}_2|) \tag{20}$$

where V is the system volume and k_B is the Boltzman constant.

We can divide the volume of the system into two parts $V = V_1 + V_2$, where V_1 is the volume of the droplet and $V_2\backslash$ is the volume of the liquid around the droplet. Then, the potential energy of the liquid can be written as follows:

$$\begin{aligned}E_p &= \tfrac{n_0^2}{2} \int\limits_{V_1+V_2} d^3R_1 \int\limits_{V_1+V_2} d^3R_2 F_2(|\mathbf{R}_1-\mathbf{R}_2|)\Phi(|\mathbf{R}_1-\mathbf{R}_2|) = \\ &= \tfrac{n_0^2}{2}\int\limits_{V_1} d^3R_1\int\limits_{V_1} d^3R_2 F_2(|\mathbf{R}_1-\mathbf{R}_2|)\Phi(|\mathbf{R}_1-\mathbf{R}_2|) + \\ &+ 2\tfrac{n_0^2}{2}\int\limits_{V_1} d^3R_1\int\limits_{V_2} d^3R_2 F_2(|\mathbf{R}_1-\mathbf{R}_2|)\Phi(|\mathbf{R}_1-\mathbf{R}_2|) + \\ &+ \tfrac{n_0^2}{2}\int\limits_{V_2} d^3R_1\int\limits_{V_2} d^3R_2 F_2(|\mathbf{R}_1-\mathbf{R}_2|)\Phi(|\mathbf{R}_1-\mathbf{R}_2|),\end{aligned} \tag{21}$$

where the first term is the bulk component of the droplet energy, the second term is the interaction energy of molecules inside the droplet with the molecules located outside, and the third term is the molecules' interaction energy in the liquid with the spherical pore. The first and the third terms are proportional to the volumes of the liquid droplet and liquid with the pore, correspondingly, and the second term is proportional to the surface area of the droplet.

Let us split the system into the droplet of radius a and the remainder of the volume (liquid with a pore of radius a). The potential energy E' of the separated system is written:

$$E' = \frac{n_0^2}{2} \int_{V_1} d^3R_1 \int_{V_1} d^3R_2 F_2(|\mathbf{R}_1 - \mathbf{R}_2|)\Phi(|\mathbf{R}_1 - \mathbf{R}_2|) + \\ + \frac{n_0^2}{2} \int_{V_2} d^3R_1 \int_{V_2} d^3R_2 F_2(|\mathbf{R}_1 - \mathbf{R}_2|)\Phi(|\mathbf{R}_1 - \mathbf{R}_2|), \quad (22)$$

where the first term is the energy of the droplet, and the second term is the energy of the liquid with a pore. The potential energy of the system separation is $E' - E_p$. The energy of the system separation is proportional to the surface area of the sphere. The specific separation energy per unit of the formed surface $S = 4\pi a^2$ can be written as:

$$\sigma = \frac{1}{2}(E' - E_p) = -\frac{n_0^2}{2} \int_{V_1} d^3R_1 \int_{V_2} d^3R_2 F_2(|\mathbf{R}_1 - \mathbf{R}_2|)\Phi(|\mathbf{R}_1 - \mathbf{R}_2|). \quad (23)$$

The specific separation energy (23) corresponds to the previously derived surface energy of the droplet and pore in Fowler's approximation [16,19]. Fowler's approximation corresponds to the "step" form of the molecules' density profiles on the boundary of the droplet and pore inside the liquid, with pair distribution functions similar to that of the homogeneous liquid.

The surface energy of the droplet can be defined as the difference between the total energy of the droplet and the energy of the homogeneous phase in the volume corresponding to the volume of the droplet divided by the surface area of the droplet. This definition is equivalent to distinguishing the volumetric and surface parts in the total energy of the droplet. For the model calculations, the pair distribution function was approximated as follows:

$$F_2(\mathbf{R}_1, \mathbf{R}_2) \cong \Theta(a - R_1)\Theta(a - R_2)F_2^{(0)}(|\mathbf{R}_1 - \mathbf{R}_2|), \quad (24)$$

where $F_2^{(0)}(|\mathbf{R}_1 - \mathbf{R}_2|)$ is the pair distribution function of atoms in a homogeneous liquid; and a is the droplet radius. Approximation (24) is similar to the Kirkwood–Buff approximation for a semi-bounded liquid with a flat interface [16,19,21].

The received expressions for the droplet energy, pore energy, and separation energy of a liquid into a droplet and a liquid with pore can be integrated in spherical coordinates. Thus, using Approximation (24), the energy of the liquid droplet can be written as:

$$E_{drop} = \frac{3}{2}Nk_BT + \frac{4}{3}\pi a^3 \varepsilon + \sigma S, \quad (25)$$

where

$$\varepsilon = 2\pi n_0^2 \int_0^\infty dR\, R^2 F_2^{(0)}(R)\Phi(R) \quad (26)$$

is the bulk density of the potential energy of a liquid; n_0 is the density of the number of particles in a homogeneous liquid; and S is the surface area of the droplet.

The surface energy of the droplet is defined as:

$$\sigma = \sigma_0 + \Delta\sigma, \quad (27)$$

where

$$\sigma_0 = -\frac{\pi n_0^2}{2} \int_0^\infty dR\, R^3 F_2^{(0)}(R)\Phi(R) \quad (28)$$

is the surface energy of the flat interface of the liquid in Fowler's approximation [16,19,21,22].

$$\Delta\sigma = \frac{\pi n_0^2}{2}\int_{2a}^{\infty} dR\, R^3 \Phi(R) F_2^{(0)}(R) + \frac{\pi n_0^2}{24a^2}\int_0^{2a} dR\, R^5 \Phi(R) F_2^{(0)}(R) - \\ - \frac{\pi n_0^2}{2} \cdot \frac{4\pi a}{3}\int_{2a}^{\infty} dR\, R^2 \Phi(R) F_2^{(0)}(R)$$
(29)

is an additional term for the surface energy of the droplet due to the surface curvature. In the boundary case of the large droplet radius values $\lim_{a\to\infty}\sigma = \sigma_0$, this corresponds to the surface energy of the liquid with a flat interface. For small values of the droplet radius, $\lim_{a\to 0}\sigma = 0$.

The model calculations of the size dependence of the surface energy of the droplet $\sigma(a)$ were performed for simple liquids using the Lennard–Jones potential and pair distribution function, which were obtained within the framework of the Wicks–Chandler–Anderson (WCA) thermodynamic perturbation theory [16,22–24]. Figure 7 shows the calculation results for the dependence $\sigma(a)$ of argon at the melting point. The atoms' pair distribution function was calculated according to the WCA procedure. The Lennard–Jones parameters were chosen as $\varepsilon = 124 K, \sigma = 3.418\overset{0}{A}$, and the droplet radius was depicted in terms of Bohr radius a_B. The asymptotic value of the surface energy at $a \to \infty$ is equal $\sigma_0 = 27.04$ erg/cm^2.

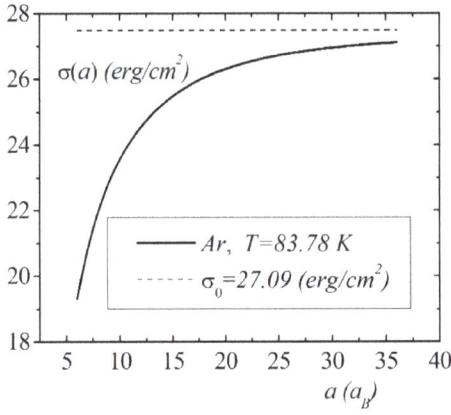

Figure 7. The size dependence of the surface energy $\sigma(a)$ for the Ar droplet.

Model calculations of the dimensional dependence of the surface energy of gases He, Kr, Xe are presented in Figures 8–10. The parameters of Lennard–Jones potentials are taken from a monograph [13]. Model calculations for all elements are performed for temperatures and densities corresponding to triple points. Note the similarity of the dimensional dependences of $\sigma(a)$ for the selected group of elements. The surface energy shows a strong dependence for nanosized droplets. With an increase in size, the surface energy of nanodroplet approaches the value of the surface energy of a flat surface σ_0.

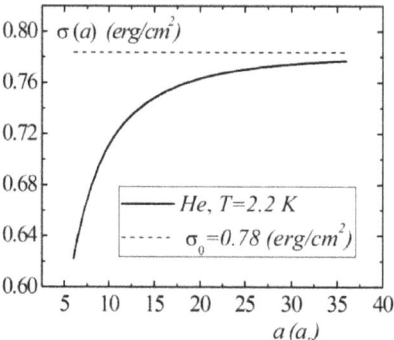

Figure 8. The size dependence of the surface energy $\sigma(a)$ for the He droplet.

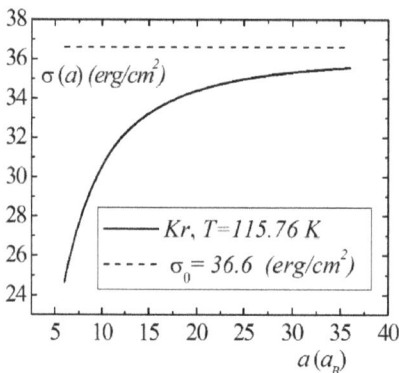

Figure 9. The size dependence of the surface energy $\sigma(a)$ for the Kr droplet.

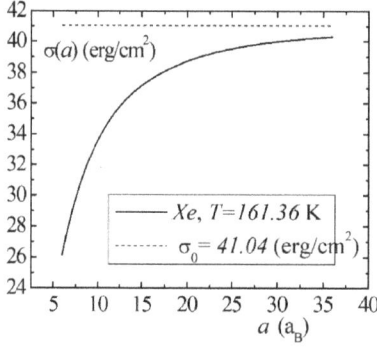

Figure 10. The size dependence of the surface energy $\sigma(a)$ for the Xe droplet.

When a droplet of liquid is located in a gas, then for the surface energy of the boundary of the droplet–gas interface, we use a representation in which the surface energy of the gas phase is described as the surface energy of a gas with a pore. We assume that the density of the number of particles in liquid is n_0, and in gas, it is n_1. The pair distribution functions of particles in a homogeneous liquid and gas are F_{20} and F_{21}, respectively. Then, the surface energy of the spherical liquid–gas interface, σ_{l-g}, is reduced in comparison to the surface energy of the liquid droplet in vacuum to a value that is equal to the surface energy of the pore with a vacuum in the gas

$$\sigma_{l-g} = \sigma_{0l} - \sigma_{0g} + \frac{\pi n_0^2}{2}\left[\int_{2a}^{\infty} dR R^3 F_{20}^{(0)}(R)\Phi(R) - \right.$$
$$\left. -\frac{4a}{3}\int_{2a}^{\infty} dR R^2 F_{20}^{(0)}(R)\Phi(R) + \frac{1}{12a^2}\int_{0}^{2a} dR R^5 \Phi(R) F_{20}^{(0)}(R)\right] - $$
$$+\frac{\pi n_1^2}{2}\left[\int_{2a}^{\infty} dR R^3 F_{21}^{(0)}(R)\Phi(R) - \frac{4a}{3}\int_{2a}^{\infty} dR R^2 F_{21}^{(0)}(R)\Phi(R) + \right.$$
$$\left. +\frac{1}{12a^2}\int_{0}^{2a} dR R^5 F_{21}^{(0)}(R)\Phi(R)\right] \quad (30)$$

where

$$\sigma_{ol} = -\frac{\pi n_0^2}{2}\int_{0}^{\infty} dR R^3 F_{20}^{(0)}(R)\Phi(R) \quad (31)$$

is the surface energy of the flat liquid–vacuum interface; and

$$\sigma_{og} = -\frac{\pi n_1^2}{2}\int_{0}^{\infty} dR R^3 F_{21}^{(0)}(R)\Phi(R) \quad (32)$$

is the surface energy of the flat gas–vacuum interface.

In the boundary case of gas density growth, $n_1 \to n_0$, the surface energy of the spherical liquid–gas interface approaches zero, $\sigma_{l-g} \to 0$. In the boundary case when $a \to \infty$, we obtain $\lim_{a \to \infty} \sigma_{l-g} = \sigma_{0l} - \sigma_{0g}$.

In this section, within the framework of the correlation theory of inhomogeneous liquids, the general expressions for the bulk and surface terms in the droplet energy as a function of the radius were obtained. Using the Fowler approximation, we were able to reduce all of the terms to the single integrals, which significantly simplified the model calculations of the size dependence. The size dependence of the surface energy was calculated for spherical droplets of the simple dielectric liquids as a function of the radius in Fowler's approximations. In the boundary case of the large radius of the droplet, the surface energy approaches the value for the flat surface. The strong size dependence of the droplet surface energy is observed at nanometer scales of the droplet radius. When the droplet radius $a < 15a_B$, a significant decrease in the surface energy is observed.

The derived approach is used to calculate the surface energy of two-phase system of nanodroplet–gas, including accounting for the paired interparticle interactions and correlations. In the vicinity of the mixing point of liquid and gas, the surface energy of the nanodroplet in gas approaches zero. The size dependence of the nanodroplet surface energy in gas is similar to the size dependence of the nanodroplet in vacuum.

4. The Correlation Theory of Interaction Energy between Two Nanodroplets and Two Nano-Pores in Liquid

The contemporary state of development in molecular–kinetic representations of the interaction of macroscopic bodies by means of intermolecular forces is described in the monograph [11]. In [11], the expressions for the interaction energy of the bodies with different geometry were derived, and these used the attractive potential of intermolecular interaction. This potential corresponds to the Van der Waals potential, and the repulsive intermolecular forces and the interparticle correlations in these calculations are not taken into account. Thus, the obtained results in [11] for the interaction energies of macroscopic bodies correspond to the asymptotic values for large distances. Taking into account the short-range intermolecular forces requires a theory that also takes into consideration the intermolecular correlations. This approach can be implemented within the framework of the statistical theory of multiparticle systems [10,13–16]. The kinetics of air dispersal and cavitation systems requires knowledge of the energy of the interaction processes of nanodroplets and nanopores as well as that in their ensembles with the surrounding phase.

In the current section, we established the theory of the interaction energy for two droplets of a simple liquid within the framework of the distribution function method of the groups of particles. The derived expressions for the interaction energy of two nanodroplets take into account paired interparticle interactions and correlations, and they are applicable for the arbitrary distances between nanodroplets or nanopores.

We postulate in a homogeneous liquid that occupies a volume V a presence of two pores of volumes V_1 and V_2. The volume of the liquid without the volume of two pores we denote V'. The potential energy of the liquid with two pores can be written:

$$E = \frac{1}{2}n_0^2 \int_{V'} d^3r_1 \int_{V'} d^3r_2 \Phi(|\mathbf{r}_1 - \mathbf{r}_2|) F_2(|\mathbf{r}_1 - \mathbf{r}_2|), \tag{33}$$

where $\Phi(|\mathbf{r}_1 - \mathbf{r}_2|)$ is the potential energy of interaction of two atoms; $F_2(|\mathbf{r}_1 - \mathbf{r}_2|)$ is the pair distribution function of atoms in a homogeneous liquid; and $n_0 = N/V$ is the density of the number of particles in a homogeneous liquid.

From the energy in (33), we can separate the part that corresponds to the energy of a homogeneous liquid in a volume V:

$$E = \frac{1}{2}n_0^2 \left(\int_{V'} d^3r_1 + \int_{V_1} d^3r_1 + \int_{V_2} d^3r_1 - \int_{V_1} d^3r_1 - \int_{V_2} d^3r_1 \right) \times$$

$$\times \left(\int_{V'} d^3r_2 + \int_{V_1} d^3r_2 + \int_{V_2} d^3r_2 - \int_{V_1} d^3r_2 - \int_{V_2} d^3r_2 \right) \Phi(|\mathbf{r}_1 - \mathbf{r}_2|) F_2(|\mathbf{r}_1 - \mathbf{r}_2|) =$$

$$= \frac{1}{2}n_0^2 \left(\int_V d^3r_1 - \int_{V_1} d^3r_1 - \int_{V_2} d^3r_1 \right) \left(\int_V d^3r_2 - \int_{V_1} d^3r_2 - \int_{V_2} d^3r_2 \right) \times$$

$$\times \Phi(|\mathbf{r}_1 - \mathbf{r}_2|) F_2(|\mathbf{r}_1 - \mathbf{r}_2|). \tag{34}$$

The energy Expression (34) contains nine terms whose physical meaning we clarify in the following. The first term is the potential energy of a homogeneous liquid in a volume V:

$$\frac{1}{2}n_0^2 \int_V d^3r_1 \int_V d^3r_2 \Phi(|\mathbf{r}_1 - \mathbf{r}_2|) F_2(|\mathbf{r}_1 - \mathbf{r}_2|) = \varepsilon V. \tag{35}$$

This energy (35) can be represented as the product of the bulk density of the potential energy ε and the volume of the system V. The sum of the contributions from (34):

$$-\frac{1}{2}n_0^2 \int_V d^3r_1 \int_{V_1} d^3r_2 \Phi(|\mathbf{r}_1 - \mathbf{r}_2|) F_2(|\mathbf{r}_1 - \mathbf{r}_2|) +$$

$$+\frac{1}{2}n_0^2 \int_{V_1} d^3r_1 \int_{V_1} d^3r_2 \Phi(|\mathbf{r}_1 - \mathbf{r}_2|) F_2(|\mathbf{r}_1 - \mathbf{r}_2|) =$$

$$= -\frac{1}{2}n_0^2 \left(\int_V d^3r_1 - \int_{V_1} d^3r_1 \right) \int_{V_1} d^3r_2 \Phi(|\mathbf{r}_1 - \mathbf{r}_2|) F_2(|\mathbf{r}_1 - \mathbf{r}_2|) = -\sigma_1 S_1, \tag{36}$$

corresponds to the potential energy of atoms' interaction inside the volume V_1 with the atoms surrounding the first pore volume $V - V_1$, which is proportional to the surface area of the first pore S_1 and the surface energy of the first pore σ_1.

The next sum of the two contributions in (34) can be written as:

$$-\tfrac{1}{2}n_0^2 \int\limits_V d^3r_1 \int\limits_{V_2} d^3r_2 \Phi(|\mathbf{r}_1-\mathbf{r}_2|)F_2(|\mathbf{r}_1-\mathbf{r}_2|)+$$
$$+\tfrac{1}{2}n_0^2 \int\limits_{V_2} d^3r_1 \int\limits_{V_2} d^3r_2 \Phi(|\mathbf{r}_1-\mathbf{r}_2|)F_2(|\mathbf{r}_1-\mathbf{r}_2|) =$$
$$= -\tfrac{1}{2}n_0^2 \left(\int\limits_V d^3r_1 - \int\limits_{V_2} d^3r_1\right) \int\limits_{V_2} d^3r_2 \Phi(|\mathbf{r}_1-\mathbf{r}_2|)F_2(|\mathbf{r}_1-\mathbf{r}_2|) = -\sigma_2 S_2,$$
(37)

which corresponds to the potential interaction energy of atoms inside the volume V_2 and atoms surrounding the second pore volume $V - V_2$. This term is proportional to the surface area of the second pore S_2 and to the surface energy of the second pore σ_2.

The sum of the following two terms:

$$+\tfrac{1}{2}n_0^2 \int\limits_{V_1} d^3r_1 \int\limits_{V_2} d^3r_2 \Phi(|\mathbf{r}_1-\mathbf{r}_2|)F_2(|\mathbf{r}_1-\mathbf{r}_2|)+$$
$$+\tfrac{1}{2}n_0^2 \int\limits_{V_2} d^3r_1 \int\limits_{V_1} d^3r_2 \Phi(|\mathbf{r}_1-\mathbf{r}_2|)F_2(|\mathbf{r}_1-\mathbf{r}_2|)$$
(38)

corresponds to the duplicated interaction energy of two droplets with volumes V_1 and V_2.

For the sum of the contributions in (34), which are not accounted for in Expressions (36)–(38), we denote ΔE and write in the form:

$$\Delta E = -\tfrac{1}{2}n_0^2 \int\limits_{V_1} d^3r_1 \int\limits_V d^3r_2 \Phi(|\mathbf{r}_1-\mathbf{r}_2|)F_2(|\mathbf{r}_1-\mathbf{r}_2|)+$$
$$+\tfrac{1}{2}n_0^2 \int\limits_{V_1} d^3r_1 \int\limits_{V_2} d^3r_2 \Phi(|\mathbf{r}_1-\mathbf{r}_2|)F_2(|\mathbf{r}_1-\mathbf{r}_2|)+$$
$$-\tfrac{1}{2}n_0^2 \int\limits_{V_2} d^3r_1 \int\limits_V d^3r_2 \Phi(|\mathbf{r}_1-\mathbf{r}_2|)F_2(|\mathbf{r}_1-\mathbf{r}_2|)+$$
$$+\tfrac{1}{2}n_0^2 \int\limits_{V_2} d^3r_1 \int\limits_{V_1} d^3r_2 \Phi(|\mathbf{r}_1-\mathbf{r}_2|)F_2(|\mathbf{r}_1-\mathbf{r}_2|) =$$
$$= -\tfrac{1}{2}n_0^2 \int\limits_{V_1} d^3r_1 \left(\int\limits_V d^3r_2 - \int\limits_{V_2} d^3r_2\right) \Phi(|\mathbf{r}_1-\mathbf{r}_2|)F_2(|\mathbf{r}_1-\mathbf{r}_2|)-$$
$$-\tfrac{1}{2}n_0^2 \int\limits_{V_2} d^3r_1 \left(\int\limits_V d^3r_2 - \int\limits_{V_1} d^3r_2\right) \Phi(|\mathbf{r}_1-\mathbf{r}_2|)F_2(|\mathbf{r}_1-\mathbf{r}_2|) =$$
$$= -\tfrac{1}{2}n_0^2 \int\limits_{V_1} d^3r_1 \int\limits_{V-V_2} d^3r_2 \Phi(|\mathbf{r}_1-\mathbf{r}_2|)F_2(|\mathbf{r}_1-\mathbf{r}_2|)-$$
$$-\tfrac{1}{2}n_0^2 \int\limits_{V_2} d^3r_1 \int\limits_{V-V_1} d^3r_2 \Phi(|\mathbf{r}_1-\mathbf{r}_2|)F_2(|\mathbf{r}_1-\mathbf{r}_2|).$$
(39)

Simplifying (39)

$$\Delta E = -\tfrac{1}{2}n_0^2 \int\limits_{V_1} d^3r_1 \left(\int\limits_{V-V_1-V_2} d^3r_2 + \int\limits_{V_1} d^3r_2\right) \Phi(|\mathbf{r}_1-\mathbf{r}_2|)F_2(|\mathbf{r}_1-\mathbf{r}_2|)-$$
$$-\tfrac{1}{2}n_0^2 \int\limits_{V_2} d^3r_1 \left(\int\limits_{V-V_1-V_2} d^3r_2 + \int\limits_{V_2} d^3r_2\right) \Phi(|\mathbf{r}_1-\mathbf{r}_2|)F_2(|\mathbf{r}_1-\mathbf{r}_2|),$$
(40)

we get

$$\Delta E = -\tfrac{1}{2}n_0^2 \int\limits_{V_1} d^3r_1 \int\limits_{V_1} d^3r_2 \Phi(|\mathbf{r}_1-\mathbf{r}_2|)F_2(|\mathbf{r}_1-\mathbf{r}_2|)-$$
$$-\tfrac{1}{2}n_0^2 \int\limits_{V_1} d^3r_1 \int\limits_{V-V_1-V_2} d^3r_2 \Phi(|\mathbf{r}_1-\mathbf{r}_2|)F_2(|\mathbf{r}_1-\mathbf{r}_2|)-$$
$$-\tfrac{1}{2}n_0^2 \int\limits_{V_2} d^3r_1 \int\limits_{V_2} d^3r_2 \Phi(|\mathbf{r}_1-\mathbf{r}_2|)F_2(|\mathbf{r}_1-\mathbf{r}_2|)-$$
$$-\tfrac{1}{2}n_0^2 \int\limits_{V-V_2-V_1} d^3r_1 \int\limits_{V_2} d^3r_2 \Phi(|\mathbf{r}_1-\mathbf{r}_2|)F_2(|\mathbf{r}_1-\mathbf{r}_2|).$$
(41)

The first term in Expression (41) corresponds to the potential energy of liquid in the volume V_1 and contains the bulk contribution $-\varepsilon V_1$. Similarly, the third term in (41) contains a bulk part $-\varepsilon V_2$ of the potential energy of a liquid in a volume V_2. Thus, (41) takes the form:

$$\Delta E = -\varepsilon \cdot V_1 - \varepsilon \cdot V_2 - \\ -\tfrac{1}{2}n_0^2 \int_{V_1} d^3r_1 \int_{V-V_1-V_2} d^3r_2 \Phi(|\mathbf{r}_1 - \mathbf{r}_2|) F_2(|\mathbf{r}_1 - \mathbf{r}_2|) - \\ -\tfrac{1}{2}n_0^2 \int_{V-V_2-V_1} d^3r_1 \int_{V_2} d^3r_2 \Phi(|\mathbf{r}_1 - \mathbf{r}_2|) F_2(|\mathbf{r}_1 - \mathbf{r}_2|). \tag{42}$$

The potential energy of the first and second droplets is $E_{1drop} = \varepsilon \cdot V_1 + \sigma_1 \cdot S_1$ and $E_{2drop} = \varepsilon \cdot V_2 + \sigma_2 \cdot S_2$, respectively.

Then, the potential energy of the liquid with two pores (33) will take the form:

$$E = \tfrac{1}{2}n_0^2 \int_{V'} d^3r_1 \int_{V'} d^3r_2 \Phi(|\mathbf{r}_1 - \mathbf{r}_2|) F_2(|\mathbf{r}_1 - \mathbf{r}_2|) = \\ = \varepsilon \cdot V + \sigma_1 \cdot S_1 + \sigma_2 \cdot S_2 - E_{1drop} - E_{2drop} + E_1 + E_2, \tag{43}$$

where:

$$E_1 = -\tfrac{1}{2}n_0^2 \int_{V_1} d^3r_1 \int_{V-V_1-V_2} d^3r_2 \Phi(|\mathbf{r}_1 - \mathbf{r}_2|) F_2(|\mathbf{r}_1 - \mathbf{r}_2|), \tag{44}$$

$$E_2 = -\tfrac{1}{2}n_0^2 \int_{V-V_2-V_1} d^3r_1 \int_{V_2} d^3r_2 \Phi(|\mathbf{r}_1 - \mathbf{r}_2|) F_2(|\mathbf{r}_1 - \mathbf{r}_2|). \tag{45}$$

The energy expressions of (44) and (45) can be rewritten as:

$$E_1 = -\tfrac{1}{2}n_0^2 \int_{V_1} d^3r_1 \int_{V-V_1} d^3r_2 \Phi(|\mathbf{r}_1 - \mathbf{r}_2|) F_2(|\mathbf{r}_1 - \mathbf{r}_2|) + \\ + \tfrac{1}{2}n_0^2 \int_{V_1} d^3r_1 \int_{V_2} d^3r_2 \Phi(|\mathbf{r}_1 - \mathbf{r}_2|) F_2(|\mathbf{r}_1 - \mathbf{r}_2|) = \\ = -\sigma_1 \cdot S_1 + \tfrac{1}{2}n_0^2 \int_{V_1} d^3r_1 \int_{V_2} d^3r_2 \Phi(|\mathbf{r}_1 - \mathbf{r}_2|) F_2(|\mathbf{r}_1 - \mathbf{r}_2|). \tag{46}$$

$$E_2 = -\tfrac{1}{2}n_0^2 \int_{V-V_2} d^3r_1 \int_{V_2} d^3r_2 \Phi(|\mathbf{r}_1 - \mathbf{r}_2|) F_2(|\mathbf{r}_1 - \mathbf{r}_2|) + \\ + \tfrac{1}{2}n_0^2 \int_{V_1} d^3r_1 \int_{V_2} d^3r_2 \Phi(|\mathbf{r}_1 - \mathbf{r}_2|) F_2(|\mathbf{r}_1 - \mathbf{r}_2|) = \\ = -\sigma_2 \cdot S_2 + \tfrac{1}{2}n_0^2 \int_{V_1} d^3r_1 \int_{V_2} d^3r_2 \Phi(|\mathbf{r}_1 - \mathbf{r}_2|) F_2(|\mathbf{r}_1 - \mathbf{r}_2|). \tag{47}$$

where the second terms in (46) and (47) correspond to the potential interaction energy of two droplets or two pores:

$$E_{drop-drop} = E_{pore-pore} = \tfrac{1}{2}n_0^2 \int_{V_1} d^3r_1 \int_{V_2} d^3r_2 \Phi(|\mathbf{r}_1 - \mathbf{r}_2|) F_2(|\mathbf{r}_1 - \mathbf{r}_2|). \tag{48}$$

Substituting Expressions (46)–(48) into (43):

$$E = \varepsilon \cdot V + \sigma_1 \cdot S_1 + \sigma_2 \cdot S_2 - E_{1drop} - E_{2drop} - \\ -\sigma_1 \cdot S_1 - \sigma_2 \cdot S_2 + E_{drop-drop} + E_{pore-pore} = \\ = \varepsilon \cdot V - E_{1drop} - E_{2drop} + E_{drop-drop} + E_{pore-pore}, \tag{49}$$

we receive the expression for the potential energy of homogeneous liquid in a volume V

$$\varepsilon \cdot V = E + E_{1drop} + E_{2drop} - E_{drop-drop} - E_{pore-pore}. \tag{50}$$

The terms in Expression (50) have the following meaning: the left part of (50) is the potential energy of a homogeneous liquid, which takes into account all interactions; the right side of (50) contains a sum that includes contributions related to the "separation" from a homogeneous liquid of two droplets with a free interface and free interface of the pores.

To derive the interaction energy of two nanodroplets $E_{drop-drop}$ and the interaction energy of two nanopores $E_{pore-pore}$ (48), we first consider the interaction energy of a volume element that contains $n_0 d^3 r_1$ atoms, with a droplet of radius a:

$$E_{el-drop}(\mathbf{R}_1) = n_0^2 dV \int_V d^3 r_2 F_2(|\mathbf{R}_1 - \mathbf{r}_2|) \Phi(|\mathbf{R}_1 - \mathbf{r}_2|) \tag{51}$$

where V is a droplet volume; and $F_2(|\mathbf{R}_1 - \mathbf{r}_2|)$ is a pair distribution function of atoms in a homogeneous liquid.

We assume that the droplet is centered in the reference coordinate system and the volume element is located along the OZ axis at distance $R_1 > a$. In a spherical coordinate system, Expression (51), similar to (5), takes the form:

$$E_{el-drop}(R_1, a) = \frac{\pi n_0 dV_1}{R_1} \Theta(R_1 - a) \int_{R_1-a}^{R_1+a} dR R \Phi(R) F_2^{(0)}(R) \times \\ \times [a^2 - R_1^2 - R^2 + 2 R R_1]. \tag{52}$$

Let us assume that the center of the second droplet with radius b is located along OZ at a point with coordinate d Then, the interaction energy of these two droplets with center distances $d > a + b$, using (52), can be written as:

$$E_{drop-drop}(a, b, d) = \int dV_1 E_{el-drop}(R_1, a). \tag{53}$$

Integrating over spherical variables (53) will take the form:

$$E_{drop-drop}(a, b, d) = \frac{\pi^2 n_0^2}{2d} \int_{d-b}^{d+b} dR_1 \left[b^2 - (d - R_1)^2 \right] \times \\ \int_{R_1-a}^{R_1+a} dR R F_2(R) \Phi(R) \left[a^2 - (R_1 - R)^2 \right]. \tag{54}$$

The Expression (54) allows calculating a component of the interaction force between two nanodroplets along the OZ direction

$$F_z(a, b, d) = -\frac{\partial}{\partial d} E_{drop-drop}(a, b, d). \tag{55}$$

A continuous density distribution is assumed in the boundary case of a continuum model, and it is not accounted for in the discreet structure of the media. The expressions derived above for the interaction energy of two nanodroplets and two nanopores can be easily reduced to the continuum case by neglecting the correlations and assuming $F_2(R) = 1$. Using the Lennard–Jones potential for the interaction energy of atoms, and integrating Expression (54) over R, we obtain:

$$E_{drop-drop}(a, b, d) = \frac{2\pi^2 n_0^2 \varepsilon \sigma^6}{d} \int_{d-b}^{d+b} dR_1 \left[b^2 - (d - R_1)^2 \right] \times \\ \{ -(a^2 \sigma^6 / 10) \left((R_1 + a)^{-10} - (R_1 - a)^{-10} \right) + (a^2/4) \left((R_1 + a)^{-4} - (R_1 - a)^{-4} \right) + \\ + (R_1^2 \sigma^6 / 10) \left((R_1 + a)^{-10} - (R_1 - a)^{-10} \right) - (2 R_1 \sigma^6 / 9) \left((R_1 + a)^{-9} - (R_1 - a)^{-9} \right) + \\ + (\sigma^6 / 8) \left((R_1 + a)^{-8} - (R_1 - a)^{-8} \right) - (R_1^2 / 4) \left((R_1 + a)^{-4} - (R_1 - a)^{-4} \right) - \\ - (1/2) \left((R_1 + a)^{-2} - (R_1 - a)^{-2} \right) + (2 R_1 / 3) \left((R_1 + a)^{-3} - (R_1 - a)^{-3} \right) \}. \tag{56}$$

Formula (56) can be simplified by integration over R_1

$$E_{drop-drop}(a,b,d) = \frac{2\pi^2 n_0^2 \varepsilon \sigma^6}{d}(b^2 - d^2) \times$$
$$\left\{ \frac{\sigma^6 a}{40} f_8^+ + \frac{11}{630} \sigma^6 f_7^- - \frac{a^2}{12} f_3^- - \frac{a^2}{12} f_3^+ + \frac{a}{3} f_2^- + \frac{a}{3} f_2^+ + \frac{1}{12} f_1^- \right\} +$$
$$+ 4\pi^2 n_0^2 \varepsilon \sigma^6 \left\{ -\frac{\sigma^6 a^3}{20} f_9^- + \frac{31}{360} a^2 \sigma^6 f_8^- + \frac{1}{90} \sigma^6 a^3 f_9^+ - \frac{1}{360} a \sigma^6 f_7^+ - \right.$$
$$\left. - \frac{1}{2160} \sigma^6 f_6^- - \frac{1}{24} a^3 f_3^+ - \frac{1}{12} a^2 f_2^- + \frac{1}{12} a f_1^- - \frac{1}{12} \ln\left|\frac{a\delta}{\gamma\beta}\right| \right\} - \qquad (57)$$
$$- \frac{2\pi^2 n_0^2 \varepsilon \sigma^6}{d} \left\{ -\frac{\sigma^6 a^3}{360} f_8^+ + \frac{\sigma^6 a^2}{168} f_7^- - \frac{\sigma^6 a}{360} f_6^+ - \frac{41\sigma^6}{1800} f_5^- - \right.$$
$$\left. - \frac{a^3}{4} f_2^+ + \frac{55 a^2}{36} f_1^- \right\}.$$

where we denoted:

$$\begin{array}{l} f_1^\pm = \alpha^{-1} - \beta^{-1} \pm \gamma^{-1} \pm \delta^{-1}, \; f_2^\pm = \alpha^{-2} - \beta^{-2} \pm \gamma^{-2} \pm \delta^{-2}, \\ f_3^\pm = \alpha^{-3} - \beta^{-3} \pm \gamma^{-3} \pm \delta^{-3}, \; f_5^\pm = \alpha^{-5} - \beta^{-5} \pm \gamma^{-5} \pm \delta^{-5}, \\ f_6^\pm = \alpha^{-6} - \beta^{-6} \pm \gamma^{-6} \pm \delta^{-6}, \; f_7^\pm = \alpha^{-7} - \beta^{-7} \pm \gamma^{-7} \pm \delta^{-7}, \\ f_8^\pm = \alpha^{-8} - \beta^{-8} \pm \gamma^{-8} \pm \delta^{-8}, \; f_9^\pm = \alpha^{-9} - \beta^{-9} \pm \gamma^{-9} \pm \delta^{-9}, \\ \alpha = d+b+a, \; \beta = d-b+a, \; \gamma = d+b-a, \; \delta = d-b-a. \end{array} \qquad (58)$$

The result in (57) corresponds to the continuum approximation for the interaction energy of two droplets of liquid with parameters $E_{drop-drop}(a,b,d)$. It accounts for attractive and repulsive parts of molecule interactions and neglects correlation effects.

The absorption problems in equilibrium liquid–vapor systems require knowledge of the nanodroplet interaction energy with the flat liquid interface. Using Expression (54) for the interaction energy of two droplets, we can investigate the boundary case of the droplet–semi-bounded liquid interaction.

These boundary case calculations of the droplet interaction energy with the semi-bounded liquid can be performed assuming that the distance between droplet centers $d \to \infty$ and droplet radius $b \to \infty$, and $d - b = const$. We denote distance $d - b = D$, which in our boundary case corresponds to the distance from the center of the droplet with radius a to the interface of semi-bounded liquid. The calculation of the limit $\lim\limits_{\substack{d \to \infty, b \to \infty \\ d-b=D}} E_{drop-drop}(a,b,d)$ is not dependent on the internal integral in (54). Thus:

$$\lim_{\substack{d \to \infty, b \to \infty \\ d-b=D}} \frac{1}{d}[b^2 - (d-R_1)^2] = 2R_1 - 2D. \qquad (59)$$

Using the result in (58), the interaction energy of the droplet of radius a with a semi-bounded liquid at a distance D will take the form:

$$E_{drop-semi}(a,b,d) = \pi^2 n_0^2 \int_D^\infty dR_1 (R_1 - D) \times$$
$$\times \int_{R_1-a}^{R_1+a} dR R F_2(R) \Phi(R)\left[a^2 - (R_1 - R)^2\right]. \qquad (60)$$

The derived Expression (60) takes into account the paired interparticle interactions by means of potential $\Phi(R)$ and the paired interparticle correlations by means of $F_2(R)$. From (59), we can derive an explicit expression for the interaction energy of the nanodroplet with a semi-bounded liquid in continuum approximation.

Figure 11 shows the results of numerical calculations of the interaction energy of two ^4He nanodroplets with the same radii $a = 2\sigma$ as a function of the distance between the centers of the droplets $4\sigma < a < 5\sigma$ in accordance with the Formula (54). The interaction energy of nanodroplets is negative and rapidly decreases with increasing distance between the centers of the droplets.

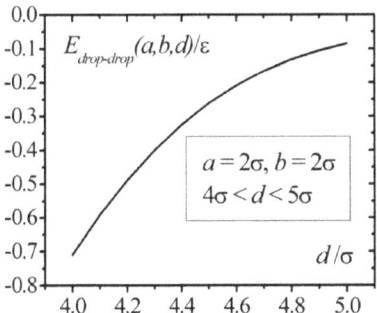

Figure 11. Dependence of the interaction energy of two ^4He nanodroplets with radii $a = 2\sigma$.

5. The Effective Hamiltonian of Aerosols with Liquid Nanodroplets

We use the derived results to build the effective Hamiltonians of nanodispersed two-phase liquid–gas systems. Our approach accounts for the effects of the paired interparticle interactions and correlations in calculations of the molecules' interaction energies with droplets, pairs of droplets, and the surface energy of the droplets.

Let us assume the existence of an aerosol of nanodroplets of different sizes in the gas phase. For simplicity, we assume that this two-phase system consists of atoms (molecules) of the same kind and has temperature T. Based on the results derived in the previous sections, the effective Hamiltonian of this system H_{eff}, including accounting for the paired interparticle interactions and correlations, can be written as:

$$H_{eff} = \tfrac{3}{2} N k_B T + K + \varepsilon_l \sum_{j=1}^{N_d} V_j + \varepsilon_g \left(V - \sum_{j=1}^{N_d} V_j \right) + \sum_{j=1}^{N_d} \sigma_j S_j + \\ + \sum_{i=1}^{N_G} \sum_{j=1}^{N_d} E_{a-d}(R_{ij}, a_j) + \tfrac{1}{2} \sum_{k=1}^{N_d} \sum_{m=1}^{N_d} E_{drop-drop}(a_k, b_m, d_{km}) \qquad (61)$$

where $\tfrac{3}{2} N k_B T$ is the kinetic energy of all molecules of aerosol; N is the total number of molecules in gas and droplets; K is the kinetic energy of the translational and rotational motion of all droplets; ε_l, ε_g are the bulk energy densities of droplets and gas (30); V is the volume of the system; V_j is the volume of the j-th droplet; N_d is the number of droplets; N_G is the number of molecules of the gas phase; σ_j, S_j is the surface energy and the surface area of the jth droplet, respectively (27); $E_{a-d}(R_{ij}, a_j)$ is the interaction energy between the i-th gas molecule and the j-th droplet (5), a_j is the radius of the j-th droplet; R_{ij} is the distance between the i-th gas molecule and the center of the j-th droplet; and $E_{drop-drop}(a_k, b_m, d_{km})$ is the interaction energy of two droplets with radiuses a_k and b_m, and a center distance $d_{km} > a_k + b_m$ (50).

The expression for the effective Hamiltonian of an aerosol (61) can be generalized for the case of multicomponent mixtures with atoms (molecules) of different kinds. Formula (61) is derived for the set of independent variables: temperature, number of atoms (molecules) in a system in total and in gas phase, size of droplets, and density of the number of atoms in liquid and gas. In real experiments with macroscopic aerosol systems, statistical datasets exist that describe the dispersion of droplet sizes and the possible disposition in external fields. Knowledge of these statistical datasets is necessary for the averaging of (61) and calculation of aerosol energy.

The prediction of aerosol behavior requires knowledge of droplets' size evolution, their concentration, the collisions results with possible coagulation, and accounting for the condensation and evaporation effects on and from the droplets' surface. The time evolution of the droplet's size distribution is described by the generalized integral–differential dynamic equation, which takes into account the balance of the number of atoms (molecules) of gas and the number of droplets, which may vary due to condensation, evaporation, and the coagulation of droplets [25–28].

The condensation phenomena, in addition to evaporation from the droplet's surface, play an essential role in many technological processes. An overview of the existing approaches to the description of the evaporation and condensation phenomena is provided in work [29]. The majority of research papers use the equations of macroscopic mechanics of continuum media and thermodynamics. However, the framework of these approaches does not allow the formulation and establishment of the boundary conditions in the droplet's near-surface area, where the application of the macroscopic equations of thermodynamics, heat transfer, and diffusion is problematic [29]. The microscopic approach developed in the current paper allows, at the molecular level, accounting for the interaction energies of atoms (molecules) with a liquid droplet, and it takes into account paired interparticle interactions and correlations with an arbitrary atom (molecule) disposition relative to the droplet. The derived expressions for $E_{a-d}(R_{ij}, a_j)$ allow research on the intermolecular forces that act on a separate particle at an arbitrary position. They also allow the development of the statistical theory of equilibrium evaporation and condensation processes for arbitrary temperatures and densities of liquid and gas in multicomponent systems with droplets of an arbitrary size. In the problems of the adhesion of liquid droplets with other molecular structures, the interaction energy $E_{a-d}(R_{ij}, a_j)$ plays an essential role and serves as a basis for further calculations.

The atom (molecule) work function from the droplet can be described as:

$$A_{a-d} = E_{a-d}(\infty, a) - E_{a-d}(0, a). \tag{62}$$

Expression (62) can be used in adhesion problems of atoms (molecules) with droplets and thermodynamics of aerosols. The sign of the atom work function of a droplet is dependent on thermodynamic conditions and may be negative. At condition $A_{a-d} > 0$, condensation processes mostly take place, whereas at thermodynamic conditions when $A_{a-d} < 0$, the evaporation effects are active. The case $A_{a-d} = 0$ corresponds to the dynamic equilibrium of the evaporation and condensation processes.

In molecular biology problems, the adhesion of large molecules (viruses) to the surface of cells and the adhesion of cells to each other play an essential role [30]. Current progress in the physics, chemistry, and experimental techniques with nanodroplets, cells, and viruses allows measurement of the adhesion forces of nanodroplets and viruses (virions) [30,31]. However, contemporary research on the energy and adhesion forces of nano-objects consists of only phenomenological developments. In this phenomenological state, the development of the molecular structures' adhesion problems and nanodroplets of liquid, or the flat liquid interface, are important for the spread of viruses in aerosols. Within the framework of the approach developed in the current article, we can write the expression for the interaction energy of the multi-atom molecule, consisting of N_{str} atoms, with a liquid droplet of radius a_j, as follows:

$$E_{drop-str} = \sum_{i=1}^{N_{str}} E_{a-d}(R_{ij}, a_j), \tag{63}$$

where R_{ij} is the distance between the i-th atom of the structure and the j-th droplet.

Discussion about the adhesion mechanism of cells and viruses is long standing. The main question is whether the adhesion occurs at the direct contact of two objects or by means of intermediate molecular structures [30,31]. The same question concerns the mechanism of adhesion of two nanodroplets. The adhesion energy of two droplets with radiuses a_1, a_2, which are in contact with each other, corresponds to the interaction energy $E_{drop-drop}(a_1, a_2, d)$, where d is the center distance between droplets.

The energy of indirect interaction of two nanodroplets with radiuses a_1, a_2 over a molecular structure with N_{str} atoms can be written:

$$E^{ind}_{drop-drop}(a_1, a_2, d) = \sum_{k=1}^{2} \sum_{i=1}^{N_{str}} E_{a-d}(R_{il}, a_j). \tag{64}$$

Formula (64) for the energy of the indirect interaction corresponds to the adhesion energy for the system of three bodies, and it can be defined as a separation work for all three components of the system to infinite distances. The boundary cases of Formula (64) describe the energy of indirect interaction of two half-spaces over the molecular structure between them. It is also important to note that all derived interaction energy Expressions (61)–(64) explicitly account for all paired interparticle interactions and correlations.

The collisions of droplets play an important role in the aerosol coagulation phenomena. The interaction energy of two droplets contains the direct interaction component $E_{drop-drop}(a_1, a_2, d)$ and the component of indirect interaction $E^{ind}_{drop-drop}(a_1, a_2, d)$ with the surroundings (atoms, molecules, molecular structures). For nanodroplet coagulation problems, one has to take into account both components. For the problems of cell and virus adhesion calculations, the comparison of direct and indirect interactions is required.

6. Discussion

The widely used continuum model of condensed systems has a limited application in the description of atoms' interactions with condensed bodies [11,12]. The model neglects the discrete atomic structure of condensed bodies, interatomic correlations, and the ability of atoms to penetrate the condensed bodies. This is clearly visible with an example of interaction of separate atoms with droplets of liquids (Figure 6). The neglect of interatomic correlations leads to an unsatisfactory description of the physics of processes that are responsible for the dynamic equilibrium of the liquid–gas system.

When calculating the surface properties of condensed systems that are different in nature, the Fowler approximation is essential, because it correctly takes into account the basic surface contributions to different thermodynamic quantities. Corrections to this approximation, due to the difference between the real density profile in the near-surface layer and the stepped layer, are of an additive nature when calculating thermodynamic functions. Going beyond the Fowler model in our problem will not change the main results and conclusions of the work. Slight changes in the energy dependence of the interaction between the atom and the nanodroplet can be expected in the near-surface region.

For condensed systems, a characteristic property is the same order of magnitude of the average kinetic and potential energies of atoms and molecules. For nano-objects, due to the large proportion of particles present in the near-surface layer, interatomic correlations can be of great importance for the formation of a self-consistent potential, which ensures the stability of the object in relation to the decay of constituent atoms or molecules. The processes of nucleation, and the mechanisms of growth and evolution of nanostructures under conditions of microscopic instability, are an integral part of the physics of phase transitions of the first kind. From the point of view of kinetics, the processes of nucleation of nano-objects take place with the participation of each individual atom, which evolves in the field of other atoms. The evolution of a single atom depends on the self-consistent field formed by other atoms and correlations with neighboring atoms. Stability of processes and their direction are also important for the growth or decay of nano-objects. The microscopic nature of the mono-atomic mechanisms of growth or decay of nanostructures, when the curvature of the surface of the new nanophase is significant, has been insufficiently studied. The solution of kinetic problems of this type is possible by means of theoretical calculations of the atom's interaction energy with the inhomogeneous environment at a nanometer scale. A proper description of the multiscale processes of nanostructure growth also requires knowledge of the energy balance of the entire nanostructure, which requires energy calculations of highly heterogeneous nanoscale systems.

The distribution functions method of groups of atoms allows the equilibrium properties of the liquid–gas interface to be expressed in terms of potentials of interatomic interactions and distribution functions of atoms. Among the most important thermodynamic functions are the surface energy and surface tension of flat and curved surfaces. When calculating the surface properties of the liquid–gas interface, the use of the Fowler approximation [18,19] for unary and binary atom distribution functions allows us to make estimates of surface contributions with considerable accuracy [10,16,19,22]. Corrections to the Fowler approximation strongly depend on the approximation methods for the non-central part of the pair distribution function and on the choice of the atomic density profile [10,16,19,22]. The main efforts in modern statistical physics relating to two-phase liquid–gas systems are aimed at calculating the atom's density profile near the phase interface and calculating its surface energy and surface tension.

In the current work, within the framework of the distribution function method of the group of particles and Fowler's approximation, we derive the expressions for the atom interaction energy with a nanodroplet of simple liquid. The derived formulation is applied for the calculations of the nanodroplets of liquid helium. The developed approach has similarities with the functional density method (FDM) [32–36]; however, it additionally takes into account the energies of the paired interparticle correlations, which are problematic in FDM. The analysis of different approaches for the accounting of pair correlation energies in the FDM in application to liquid helium is described in paper [36]. Assembling the expression for the energy of pair correlations only by means of the atom's paired interaction energy and local densities of the number of atoms requires the introduction of the non-physical "screening" concept of Lennard–Jones potential at small distances, to avoid the divergence problem of the corresponding integrals [36]. In fact, it is necessary to introduce additional parameters that cannot be determined within the density functional method itself. As a result, extra designations for Lennard–Jones potential are introduced in different versions of FDM at short distances in an attempt to converge the integrals corresponding to the indirect accounting for the short-range paired interatomic correlations. The distribution function method used above for groups of particles avoids the necessity of expanding the studied quantities into gradient series.

The model calculations performed above for the size dependence of the atoms' interaction with a nanodroplet of helium indicates a significant influence of the saturation effect of the single atomic energy inside the droplet. The short range of interatomic interactions results in the rapid achievement of the asymptotic value of the atoms' energy at the center of the droplet during the growth of the droplet size. The formation of the bulk properties of helium nanodroplets occurs at radii in the order of 4σ and the number atoms of around 101. It should be noted that the pair distribution function of atoms in liquids at distances 4σ is almost equal to one, which corresponds to the absence of correlations in the spatial positions of the pair of atoms. Therefore, it is possible to reach conclusions about the relationship between the radii of liquid nanodroplets, at which the volumetric properties of the liquid are formed, and the characteristic distances at which the paired interatomic correlation in liquids is lost. This conclusion also easy to see in Figure 5, from which it follows that most atoms in droplets of the specified size have the same one-atomic potential as in a homogeneous liquid. The saturation effect is also observed for the thickness of the near-surface region, in which the monoatomic potential changes its value from its "bulk" value to zero outside the droplet. The depth of the potential well, in which the atom moves inside the helium droplet, for the droplets with radii of the order $a > 4\sigma$, has a magnitude $5\varepsilon \approx 50K$, which significantly exceeds the thermal energy of the atoms ($T = 2.2K$).

The kinetic energy of helium atoms contains the classical thermal contribution and the quantum contribution. Calculations of the kinetic energy of helium atoms in a wide range of densities and temperatures using the Monte Carlo quantum method [35,36] indicate the importance of accounting for the quantum contributions. At the selected temperature and density of liquid helium, the kinetic energy, according to the results in [36], is of the order $15K$, which means that the main contribution to the kinetic energy is made by quantum effects. A comparison of the kinetic energy with the depth of the potential well shows that the atoms are localized in the droplet and that helium nanodroplets tend to be resistant to spontaneous decay due to thermal fluctuations and the emission of individual atoms.

7. Conclusions

In this paper, in the framework of the correlation theory of inhomogeneous liquids, general expressions are obtained for the volume and surface contributions to the energy of a droplet as a function of its radius. In Fowler's approximation, all contributions are expressed in the form of one-time integrals, which significantly simplifies the model calculations of the dimensional dependence. The dimensional dependence of the surface energy of spherical droplets of simple dielectric liquids as a function of the radius in the Fowler approximation is calculated. In the extreme case of large values of the radius of a droplet, its surface energy approaches its value for a flat surface. The strong dimensional dependence of the surface energy of the droplet is observed in the region of nanometer droplet size. The developed approach is applied to the calculation of the surface energy of a two-phase system of liquid nanodroplet–gas phase, taking into account paired interparticle interactions and correlations. The dimensional dependence of the surface energy of a nanodroplet in a gas is similar to the dimensional dependence of a nanodroplet in a vacuum.

This paper shows the possibility of taking into account paired interparticle interactions and correlation effects when calculating the interaction energy of pairs of droplets of arbitrary size. The energy of the interaction of two droplets in the Fowler approximation is written as a double integral. An explicit expression for the interaction energy of two droplets in the continuum approximation is obtained, which allows us to investigate the importance of taking into account the effects of paired intermolecular correlations in comparison with continuum models. As a boundary case, the expression for the energy of interaction of a droplet and a semi-bounded liquid is obtained, taking into account the effects of paired interparticle correlations.

On the basis of the constructed theory, the properties of nanometer air dispersal systems with arbitrary droplet dispersion and for arbitrary multicomponent mixtures of liquids can be calculated. By means of the effective Hamiltonian of the aerosol of liquid droplets (61), it is possible to investigate in more detail the kinetic problems of evaporation and condensation on the surfaces of droplets and the lifetime of nanodroplets during their evaporation.

Within the framework of the developed approach, it is possible to study the interaction of liquid nanodroplets with foreign molecular structures (e.g., virions), coagulation of molecular structure and nanodroplets, and evaporation and condensation processes on droplets, provided that the molecular structure is present inside the droplet. These issues are all currently unresolved. The important theoretically unresolved issues are the evolution processes of evaporation and condensation on a nanodroplet that is in contact with different media (wood, plastic, metal), and the lifetime of the droplet with a molecular structure inside different surfaces. The answers to these questions are important in the problems related to the spread of viruses.

Funding: This research received no external funding.

Data Availability Statement: Data available in a publicly accessible repository.

Conflicts of Interest: The author declare no conflict of interest.

References

1. Morgenstern, J. Aerosols, Droplets, and Airborne Spread: Everything You Could Possibly Want to Know. *First10EM Blog*, 6 April 2020. Available online: https://first10em.com/aerosols-droplets-and-airborne-spread/ (accessed on 23 December 2020).
2. Miller, S. Coronavirus Drifts through the Air in Microscopic Droplets—Here's the Science of Infectious *Aerosols*, 24 April 2020. Available online: https://theconversation.com/coronavirus-drifts-through-the-air-in-microscopic-droplets-heres-the-science-of-infectious-aerosols-136663 (accessed on 23 December 2020).
3. Anfinrud, P.; Stadnytskyi, V.; Bax, C.E.; Bax, A. Visualizing Speech-Generated Oral Fluid Droplets with Laser Light Scattering. *N. Engl. J. Med.* **2020**, *382*, 2061–2062. [CrossRef] [PubMed]
4. Asadi, S.; Wexler, A.S.; Cappa, C.D.; Barreda, S.; Bouvier, N.M.; Ristenpart, W.D. Aerosol emission and superemission during human speech increase with voice loudness. *Sci. Rep.* **2019**, *9*, 2348. [CrossRef] [PubMed]
5. Hsiao, T.C.; Chuang, H.C.; Griffith, S.M.; Chen, S.J.; Young, L.H. COVID-19: An Aerosol's Point of View from Expiration to Transmission to Viral-mechanism. *Aerosol Air Qual. Res.* **2020**, *20*, 905–910. [CrossRef]
6. Meyerowitz, E.A.; Richterman, A.; Gandhi, R.T.; Sax, P.E. Transmission of SARS-CoV-2: A Review of Viral, Host, and Environmental Factors. *Ann. Intern. Med.* **2020**. [CrossRef] [PubMed]
7. World Health Organization (WHO). *Infection Prevention and Control of Epidemic-and Pandemic Prone Acute Respiratory Infections in Health Care: WHO Guidelines*; WHO Press: Geneva, Switzerland, 2014.
8. Vanag, V.K.; Epstein, I.R. Patterns of Nanodroplets: The Belousov-Zhabotinsky-Aerosol OT-Microemulsion System. In *Self-Organized Morphology in Nanostructured Materials*; Al-Shamery, K., Parisi, J., Eds.; Springer: Berlin/Heidelberg, Germany, 2008; Volume 99, pp. 89–113.
9. Lussier, D.T.; Kakalis, N.M.P.; Ventikos, Y. Molecular Dynamics Modeling of Nanodroplets and Nanoparticles. In *Multiscale Modeling of Particle Interactions: Applications in Biology and Nanotechnology*; King, M.R., Gee, D.J., Eds.; John Wiley & Sons, Inc.: Hoboken, NJ, USA, 2010; pp. 151–183. [CrossRef]
10. Rowlinson, J.S.; Widom, B. *Molecular Theory of Capillarity*; Clarendon Press: Oxford, UK, 1982.
11. Israelachvili, J.N. *Intermolecular and Surface Forces*, 3rd ed.; Academic Press: Cambridge, MA, USA, 2015.
12. Bruch, L.W.; Cole, M.W.; Zaremba, E. *Physical Adsorption: Forces and Phenomena*; Clarendon Press: Oxford, UK, 1997.
13. Hirschfelder, J.O.; Kurtiss, C.F.; Bird, R.B. *Molecular Theory of Gases and Liquids*; John Wiley and Sons Inc.: New York, NY, USA, 1954.
14. Fisher, I.Z. *Statistical Theory of Liquids*; The University of Chicago Press: Chicago, IL, USA, 1964.
15. Balesku, R. *Equilibrium and Non-Equilibrium Statistical Mechanics*; Wiley-Interscience: New York, NY, USA, 1975.
16. Croxton, C.A. *Liquid State Physics—A Statistical Mechanical Introduction*; Cambridge University Press: Cambridge, UK, 1974.
17. Giessibl, F.J. Advances in atomic force microscopy. *Rev. Mod. Phys.* **2003**, *75*, 949–983. [CrossRef]
18. Fowler, R.H. A Tentative Statistical Theory of Macleod's Equation for Surface Tension and the Parachor. *Proc. R. Soc.* **1937**, *159*, 229–246.
19. Ono, S.; Kondo, S. Molecular Theory of Surface Tension in Liquids. In *Structure of Liquids*; Encyclopedia of Physics X; Flügge, S., Ed.; Springer: Berlin, Germany, 1960.
20. Landau, L.D.; Lifshitz, E.M. *Statistical Physics*; Butterworth-Heinemann: Oxford, UK, 1980; Volume 5.
21. Kirkwood, G.; Buff, F.P. The statistical mechanical theory of surface tension. *J. Chem. Phys.* **1949**, *17*, 338–343. [CrossRef]
22. March, N.H.; Tosi, M.P. *Atomic Dynamics in Liquids*; Dover Publications: New York, NY, USA, 1991.
23. Verlet, L.; Weis, J.J. Equilibrium theory of simple liquids. *Phys. Rev. A* **1972**, *5*, 939–952. [CrossRef]
24. Weeks, J.D.; Chandler, D.; Andersen, H.C. Perturbation theory of the thermodynamic properties of simple liquids. *J. Chem. Phys.* **1971**, *55*, 5422–5423. [CrossRef]
25. Friedlander, S.K. *Smoke, Dust and Haze: Fundamentals of Aerosol Behavior*; Oxford University Press: Oxford, UK, 2000.
26. Williams, M.M.R.; Loyalka, S.K. *Aerosol Science: Theory and Practice: With Special Applications to the Nuclear Industry*; Pergamon Press: Oxford, UK, 1991.
27. Whitby, E.R.; McMurry, P.H. Modal Aerosol Dynamics Modeling. *Aerosol Sci. Technol.* **1997**, *27*, 673–688. [CrossRef]
28. Lushnikov, A.A. Introduction to Aerosols. In *Aerosols—Science and Technology*; Agranovski, I., Ed.; WILEY-VCH Verlag GmbH & Co.: Weinheim, Germany, 2010; pp. 1–41.
29. Sazhin, S.S. Classical and Novel Approaches to Modelling Droplet Heating and Evaporation. In *Droplet Interactions and Spray Processes*; Lamanna, G., Tonini, S., Cossali, G.E., Weigand, B., Eds.; Springer Nature Switzerland AG: Cham, Switzerland, 2020; pp. 251–258. [CrossRef]
30. Kendall, K.; Kendall, M.; Rehfeldt, F. *Adhesion of Cells, Viruses, and Nanoparticles*; Springer: Heidelberg, Germany, 2011. [CrossRef]
31. Wiegand, T.; Fratini, M.; Frey, F.; Yserentant, K.; Liu, Y.; Weber, E.; Galior, K.; Ohmes, J.; Braun, F.; Herten, D.-P.; et al. Forces during cellular uptake of viruses and nanoparticles at the ventral side. *Nat. Commun.* **2020**, *11*, 1–13. [CrossRef] [PubMed]
32. Stringari, S.; Treiner, J. Systematics of liquid helium clusters. *J. Chem. Phys.* **1987**, *87*, 5021. [CrossRef]
33. Dupont-Roc, J.; Himbert, M.; Pavloff, N.; Treiner, J. Inhomogeneous liquid 4He: A density functional approach with a finite-range interaction. *J. Low Temp. Phys.* **1990**, *81*, 31–44. [CrossRef]
34. Dalfovo, F.; Lastri, A.; Pricaupenko, L.; Stringari, S.; Treiner, J. Structural and dynamical properties of superfluid helium: A density-functional approach. *Phys. Rev. B* **1995**, *52*, 1193. [CrossRef] [PubMed]

35. Szybisz, L. Comparison of density-functional approaches and Monte Carlo simulations for free planar films of liquid. *Eur. Phys. J. B* **2000**, *14*, 733–746. [CrossRef]
36. Ceperley, D.M.; Simmons, R.O.; Blasdell, R.C. The Kinetic Energy of Liquid and Solid 4He. *Phys. Rev. Lett.* **1996**, *77*, 115–118. [CrossRef] [PubMed]

Article
Self-Assembled Structures of Colloidal Dimers and Disks on a Spherical Surface

Nkosinathi Dlamini [1,†], **Santi Prestipino** [2,†] **and Giuseppe Pellicane** [1,3,4,*,†]

1. School of Chemistry and Physics, University of Kwazulu-Natal and National Institute of Theoretical Physics (NIThEP), Pietermaritzburg 3209, South Africa; Dlamini2@ukzn.ac.za
2. Dipartimento di Scienze Matematiche ed Informatiche, Scienze Fisiche e Scienze della Terra, Università degli Studi di Messina, Viale F. Stagno d'Alcontres 31, 98166 Messina, Italy; sprestipino@unime.it
3. Dipartimento di Scienze Biomediche, Odontoiatriche e delle Immagini Morfologiche e Funzionali, Università degli Studi di Messina, 98125 Messina, Italy
4. CNR-IPCF, Viale F. Stagno d'Alcontres, 98158 Messina, Italy
* Correspondence: gpellicane@unime.it
† The authors contributed equally to this work.

Abstract: We study self-assembly on a spherical surface of a model for a binary mixture of amphiphilic dimers in the presence of guest particles via Monte Carlo (MC) computer simulation. All particles had a hard core, but one monomer of the dimer also interacted with the guest particle by means of a short-range attractive potential. We observed the formation of aggregates of various shapes as a function of the composition of the mixture and of the size of guest particles. Our MC simulations are a further step towards a microscopic understanding of experiments on colloidal aggregation over curved surfaces, such as oil droplets.

Keywords: molecular self-assembly; amphiphilic aggregates; spherical boundary conditions

1. Introduction

Colloidal particles dispersed in a fluid medium are widely considered to be an ideal system where self-assembly can be explored, since they can be resolved and tracked in real time using optical microscopy [1]. The aggregation of colloidal particles is often the outcome of steric stabilization by electrostatic repulsion, which is achieved by modifying the salt concentration or by adding chemicals as stabilizing agents, as in the case of gold colloids [2] or silica and polystyrene particles [3,4]. The morphology of colloidal aggregates depends on the prevailing aggregation mechanisms and on particle shape [1], and is typically observed in ramified or compact clusters of fractal dimensionality [5], in a number of crystalline and amorphous solids [6], and in mesophases [7–9], which are partially ordered phases that are intermediate between liquids and crystals (e.g., cluster fluids, liquid crystals, and quasicrystals). Colloidal nanocrystals are even able to self-assemble in crystalline superlattices with an intricate structure [10].

In the last few decades, many researchers focused on the self-assembly of colloidal particles at an interface, which may serve as a scaffold or template for particle aggregation. Assembly at air–liquid and liquid–liquid interfaces is driven by a complex interplay of entropic and enthalpic forces [11]. The ability of oil–water interfaces to trap micron-sized particles has been known for over a century [12,13], and the strong binding of colloidal particles to fluid interfaces (the binding energy is even thousand times stronger than the thermal energy) is also evidenced by the stabilization of foams and emulsions against decomposition [14,15]. The self-assembly of colloidal particles on a flat surface can only rely on the control of interparticle interactions at the interface [16–20]. However, thanks to recent progress in microfluidics, it is possible to modify the interfacial geometry to trap colloidal particles, thus extending the initial range of applications of colloidal self-assembly. Indeed, curved phases of matter are found in a large number of systems, including biological entities such as cells and viral capsids, and the competition between the tendency to self-assemble and the geometric frustration originated by the curved interface can lead to novel structures, which are simply impossible to obtain over flat interfaces [21]). Recently, spherical boundary conditions have begun to also be employed in the realm of ultracold quantum particles, where "phases" with polyhedral symmetry are expected [22], and Bose–Einstein condensation has peculiarities that are experimentally well within reach [23].

Concrete realizations of spherical crystals are found in the emulsions of two immiscible fluids, such as oil and water, which are stabilized against droplet coalescence by coating the interface of one of the fluids with small colloidal particles [24]. A nontrivial issue is overcoming the strong binding of colloidal particles to the (liquid–liquid) interface and allowing them to diffuse quickly enough, i.e., like in a true fluid, to facilitate self-assembly over the substrate. Recently, that was achieved by very efficient functionalization with complementary DNA strands of both the surface of oil droplets (which was stabilized with sodium dodecyl sulphate (SDS), i.e., a micelle-forming surfactant) and the surface of colloidal particles [25]. Fluidlike diffusion was reached by allowing colloidal particles to anchor on rafts of polylysine-g[3.5]-polyethylene glycol-biotin (PLL-PEG-bio), which are free to slide on the surfactant. Upon increasing the concentration of SDS, the colloidal particles attached to the surface were observed to undergo aggregation as a result of the depletion effect driven by the excluded area of the surfactant micelles [25].

Spherical droplets were also coated with polystyrene latex particles [26] to form aggregates with a rigid shell, called colloidosomes in analogy to liposomes [27]. Structures resulting from the encapsulation of colloidal particles [28] are potential candidates for the delivery of drugs and vaccines, and may be used as vehicles for the slow release of cosmetic and food supplements.

In general, the self-assembly of colloidal particles on a spherical surface is an important paradigm to understand the structuring of membrane cells, which exhibit stable domains. The distribution and composition of these domains over the cell surface determines the interaction energy between different cells, ultimately driving the organization of the crowded environment inside biological organisms, where a huge number of cells is present [29].

Recently, we studied in 3D space [30] and on a plane [31] an implicit-solvent description of the dispersion of two colloidal species, namely, an amphiphilic dimer and a guest spherical/circular particle, where the smaller monomer in the dimer was solvophobic and had a strong affinity for the guest particle. In this paper, we consider the same system embedded in a spherical surface. By establishing bonds with two nearby curved disks, the smaller monomer provides the glue that keeps the disks together, which is the mechanism by which disks can form aggregates. However, once an aggregate of disks is covered with dimers, further growth of the aggregate is obstructed by the steric hindrance of the coating shell. Since the dimer–disk attraction is of limited range, mostly zero ("micelles")- and one-dimensional aggregates ("chains") are expected to form for a moderately low number of disks, while two-dimensional self-assembled structures (i.e., stratified lamellae) could occur under equimolar conditions.

The paper is organized as follows. In Section 2, we describe the model and employed method. In Section 3, we present and discuss our results. Lastly, we report our conclusions in Section 4.

2. Model and Method

The investigated model is the curved-surface analog of the same mixture of dimers and spherical guest particles that was studied in [30–32]. A dimer consists of two tangent hard calottes (i.e., disks following the surface of the sphere) with curved diameters σ_1 and $\sigma_2 = 3\sigma_1$, whereas guest particles are represented as hard calottes of size σ_3 (below, we generically refer to these particles as "disks"). In addition to the impenetrability of all particle cores, we added an attraction between the small monomer and the disk, modeled as a square-well potential of depth ϵ; the width of the well was set to be equal to σ_1. In the following, σ_2 (i.e., the diameter of the large monomer) and ϵ are taken as units of length and energy. Lastly, $N_1 = N_2$ and N_3 are the number of dimers and disks, respectively; hence, $N = N_1 + N_3$ is the total number of particles and $\chi = N_3/N$ is the (disk) composition.

Most data were collected for a fixed number $N_3 = 400$ of disks with diameter $\sigma_3 = \sigma_2$ and varying composition. Number density was $\rho^* \equiv (N/A)\sigma_2^2 = 0.05$ (with A being the area of the spherical surface), but we also performed a few simulations for $\rho^* = 0.25$ to probe the regime of moderately high curvatures. We analyzed the system behavior for a number of compositions: $\chi = 20\%, 33\%, 50\%$, and 80%. Once N and χ are set, the number of dimers follows accordingly (sphere radius R is uniquely determined from N and ρ^*). We also examined how self-assembly changes when the disk diameter is increased up to $5\sigma_2$.

Simulations were carried out using the standard Metropolis algorithm in the canonical ensemble. Typically, a few hundred million Monte Carlo (MC) cycles are performed, one cycle consisting of N trial moves. Both translational and rotational moves are carried out for dimers. In performing a translational move, the midpoint of the arc joining the monomer centers is randomly shifted on the sphere, while keeping the direction of the subtended chord fixed in the embedding three-dimensional space. In a rotational move, it is the midpoint of the arc between the monomers that is fixed, while the chord is rotated at random. The maximal random shift and rotation were adjusted during the equilibration run, so as to keep the ratio of accepted to total number of moves close to 50%. The schedule of each move was designed so that the detailed balance held exactly. Particles are initially distributed at random on the sphere (using a variant of the Box–Muller algorithm [33]). Then, the system is quenched to $T^* \equiv k_BT = 0.15$ or 0.10 and subsequently relaxed until some stationary condition is established. We checked that a slow cooling of the system, starting at each temperature from the last configuration produced at a slightly higher temperature, did not make any substantial difference in the structural properties of the steady state, because simulated systems are overall dilute.

A property signalling how far the system is away from equilibrium is the total potential energy U: an energy fluctuating around a fixed value for long is the hallmark of (meta)stable equilibrium. In ϵ units, U gives the total number of 1–3 contacts in the current system configuration. Hence, a stationary value of U indicates that aggregates eventually reached a nearly stable structure. Typically, 10^8 cycles suffice for reaching a stationary state of low density. This is clearly illustrated in Figure 1, showing the energy evolution as a function of Monte Carlo cycles for $T^* = 0.10, \rho^* = 0.05$, and a number of compositions. Once equilibrium (or whichever steady state) is reasonably attained, we gain insight into the nature of aggregates mainly by visual inspection. We also computed the radial distribution function (RDF) of disks, $g_{33}(r)$ in a rather long production run of 10^7 cycles (we checked that statistical errors on the RDFs were indeed negligible). For the sake of comparison, similar studies of 2D binary mixtures at considerably higher total density were executed for one order of magnitude fewer MC steps [34]. Even in a strongly heterogeneous system where mesoscopic structures were present, $g_{33}(r)$ bears valuable information on the arrangement of disks in the close neighborhood of a reference disk. Two disks form a bound pair when their distance is not larger than $r_{\min} = \sigma_3 + 3\sigma_1$ [31]: this is the maximal distance at which two disks can still be in contact with the same small monomer (exactly placed in the middle). For $\sigma_3 = \sigma_2$, this implies that a disk forms bonds with all its first and second neighbors, identified as such through the RDF profile.

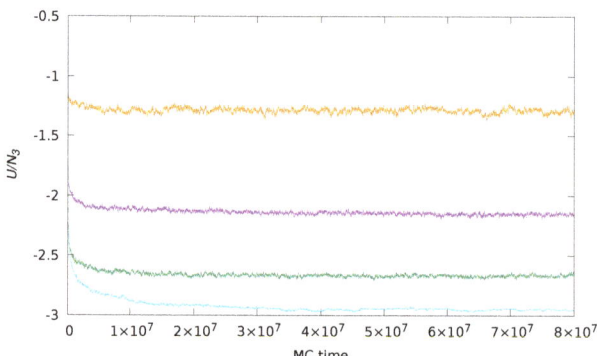

Figure 1. Energy evolution as a function of Monte Carlo time for $\sigma_3 = \sigma_2$, $T^* = 0.10$, $\rho^* = 0.05$, and various disk compositions (from top to bottom, $\chi = 20\%, 33\%, 50\%,$ and 80%).

3. Results

We first comment on the simulation results for a mixture of disks and large monomers having the same size (Section 3.1). We consider systems of both low density ($\rho^* = 0.05$) and moderate density ($\rho^* = 0.25$). Next, we examine what changes when dimers are much smaller than the disks (Section 3.2). By visual inspection, we could easily ascertain the nature of the structures present in the stationary configurations of the low-temperature system.

3.1. Same Size of Dimers and Disks

We initially set the density to be equal to $\rho^* = 0.05$, and disk size to $\sigma_3 = \sigma_2$. For $T^* = 0.15$, the equilibrated system is a fluid of small globular clusters for all disk compositions; see examples in Figure 2. Only for values of χ lower than about 10% did dimers form a well-definite coating shell around disks. Thermal fluctuations for $T^* = 0.15$ are still too important to allow for the formation of more elaborate structures.

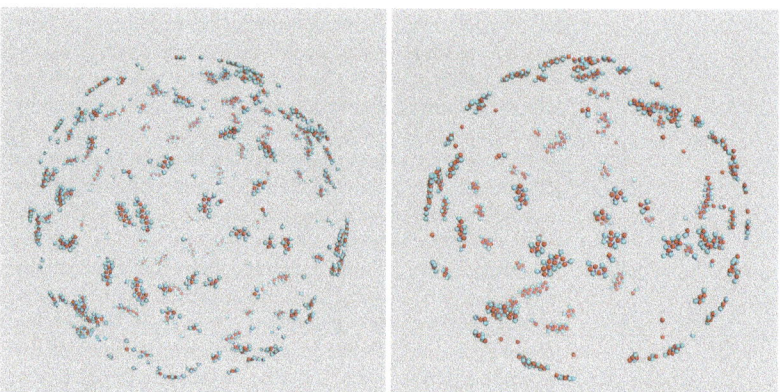

Figure 2. Typical configuration of mixture for $\sigma_3 = \sigma_2$ and $\rho^* = 0.05$ after a long run at $T^* = 0.15$. Snapshots refer to a system of composition (left) $\chi = 33\%$ and (right) $\chi = 50\%$.

Things changed radically for $T^* = 0.10$, where the nature of aggregates was more varied. For $\chi = 20\%$ or lower, we invariably observed small groups of disks surrounded by dimers (Figure 3, top-left panel); for higher compositions up to 50%, aggregates were more elongated and wormlike (see top-right and bottom-left in Figure 3). A closer look at such "worms", which are obviously the 2D analog of lamellae, revealed that they were assembled from a repeating unit, like a polymer chain. For still higher χ, the mean size of aggregates returned to being small again, since the number of gluing dimers was insufficient for all disks, and many disks then remained unbound (Figure 3, bottom-right panel). Therefore, aggregates only achieved large sizes when the number of disks roughly matched that of dimers.

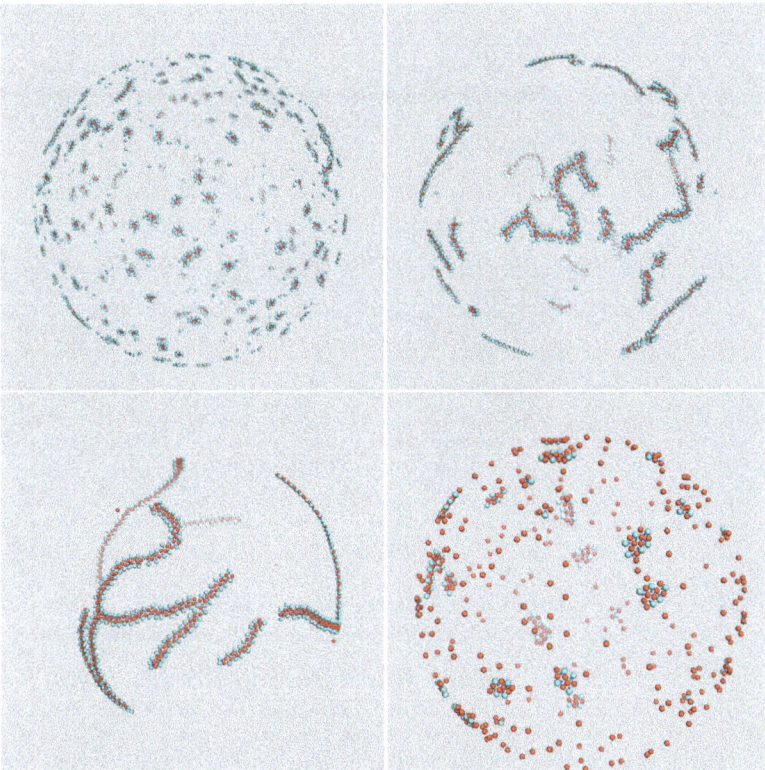

Figure 3. Typical configuration of mixture for $\sigma_3 = \sigma_2$ and $\rho^* = 0.05$ after a long run at $T^* = 0.10$. Snapshots were taken at different compositions: (from top left to bottom right) $\chi = 20\%, 33\%, 50\%$, and $\chi = 80\%$.

The dynamics of aggregation in the present model is easy to explain. Initially, when the system was still disordered, the aggregation of disks proceeded very fast through the formation of bonds between disks and dimers. As an aggregate grows in size, however, its surface becomes increasingly rich in large monomers, which are inert particles; eventually, an aggregate stops growing when its disks and small monomers all lie buried under the surface. While local adjustments of the structure still occur at a high rate, the merging of two disconnected aggregates (or the breaking of a long chain) is highly suppressed and only takes place on much longer time scales. The existence of two regimes of aggregate growth (fast and slow), corresponding to a transition from diffusion-limited to reaction-limited aggregation [3], is reflected in the crossover of U from an exponential to a subexponential decay (as evidenced in Figure 1).

For $T^* = 0.10$, the mechanism underlying the structure of aggregates is mainly energy minimization, whereas entropy considerations play a minor role. However, entropy is decisive in shaping the large-scale distribution of aggregates on the sphere, inasmuch as their structure maintains a certain flexibility (see more below). The relative size of particles and the range of 1–3 attraction are also clearly important. It is the short-range character of the attraction that is responsible for the essentially one-dimensional geometry of the larger aggregates.

Particularly interesting are the systems with $\chi = 33\%$ and $\chi = 50\%$, which are shown enlarged in Figure 4. Here, most of the aggregates are flexible worms, i.e., chainlike aggregates with a small bending modulus. The geometry of the worm backbone is dictated by the necessity to keep the energy as small as possible at the given composition: this was accomplished by a straight chain of disks for $\chi = 33\%$ (Figure 4, left panel) and by a zig-zag chain for $\chi = 50\%$ (Figure 4, right panel). Both chain morphologies allowed for disks to bind all dimers, so that, in the long run, no free particles would be left in the box. Occasionally, a worm bent to the point that a closed loop appears—see the example on the left in Figure 4.

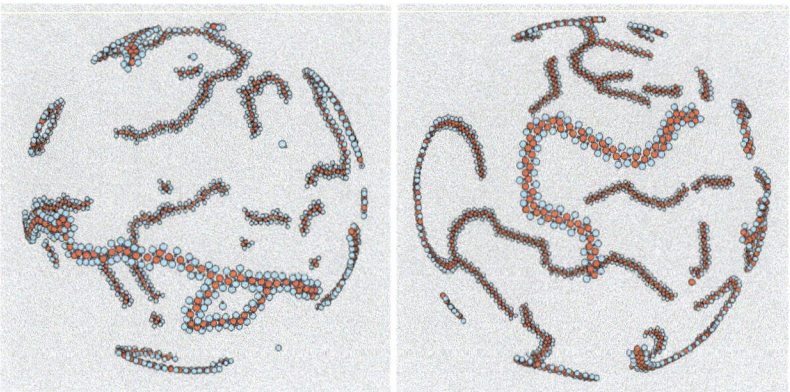

Figure 4. Typical configurations of mixture for $\sigma_3 = \sigma_2$ after "only" 3×10^7 MC cycles. Snapshots were taken for $T^* = 0.10$ and $\rho^* = 0.05$, and refer to (left) $\chi = 33\%$ and (right) $\chi = 50\%$.

Figure 5 shows the collected RDFs for various compositions at low temperature ($T^* = 0.10$). A large g_{33} value at contact is the most distinct signature of the existence of aggregates of disks. The short-distance structure in g_{33} was richer for intermediate χ values, where the physiognomy of aggregates is better defined. The rather pronounced second-neighbor peak at $\chi = 50\%$ was the result of the zig-zag structure of the chain backbone at this composition. Regarding $g_{13}(r)$, its short-distance profile was sharper for the lowest compositions, where the number of dimers that were in close contact with the same sphere was higher (see inset of right panel). A high third-neighbor peak for intermediate compositions is the signature of the existence of extended aggregates of dimers and guest disks consisting of a periodically repeating unit.

As density increases, the nature of self-assembly becomes slightly different. We studied mixtures for $\rho^* = 0.25$ while still keeping $\sigma_3 = \sigma_2$ and the temperature fixed at 0.10 (see Figure 6). When the composition was low, the aggregates were slightly elongated capsules (see top-left panel of Figure 3), like at small density. For $\chi = 33\%$, aggregates were definitely chainlike (Figure 6, top-right panel), but due to a more crowded environment, they were joined together in an intricate manner, giving rise to a spanning cluster (i.e., a connected gel-like network encompassing all particles in the system). The onset of an extended network is a remarkable outcome, considering that this structure emerged from very basic interaction rules in a binary system of disks and dimers. At the composition of 50%, large monomers were more effective in screening a chain from other aggregates, and chains then grew much longer. As a result, a chain may wrap a few times around the sphere before merging into another chain (see bottom-left panel of Figure 6). This is similar to what was observed in a one-component system of particles interacting through a short-range attractive, long-range repulsive (SALR) potential [35]. The latter system was stripe-forming at low temperature; hence, when the particles were constrained to a spherical surface, a stripe may wrap around the sphere, as indeed observed. Lastly, at still higher compositions of disks, the size of aggregates was again reduced since there were few disks to bind all dimers, and their environment was too crowded to allow for aggregates to grow long in the early stages of equilibration.

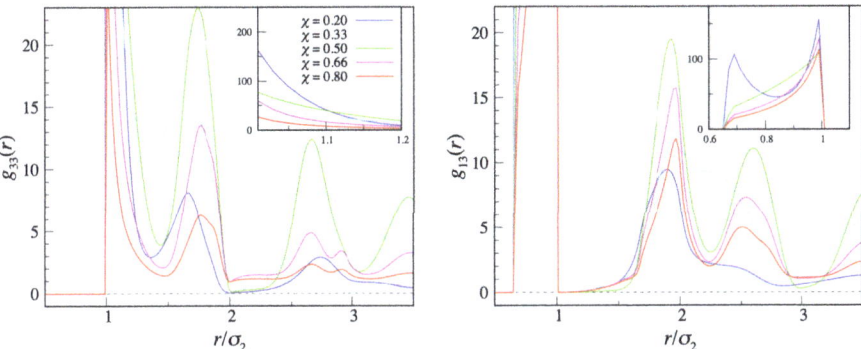

Figure 5. Mixture of dimers and disks with $\sigma_3 = \sigma_2$ after 10^8 MC cycles for $T^* = 0.10$ and $\rho^* = 0.05$. (**left**) $g_{33}(r)$; (**right**) $g_{13}(r)$; (inset) short-range structure of RDFs. The color code is the same for both panels (see left-panel inset).

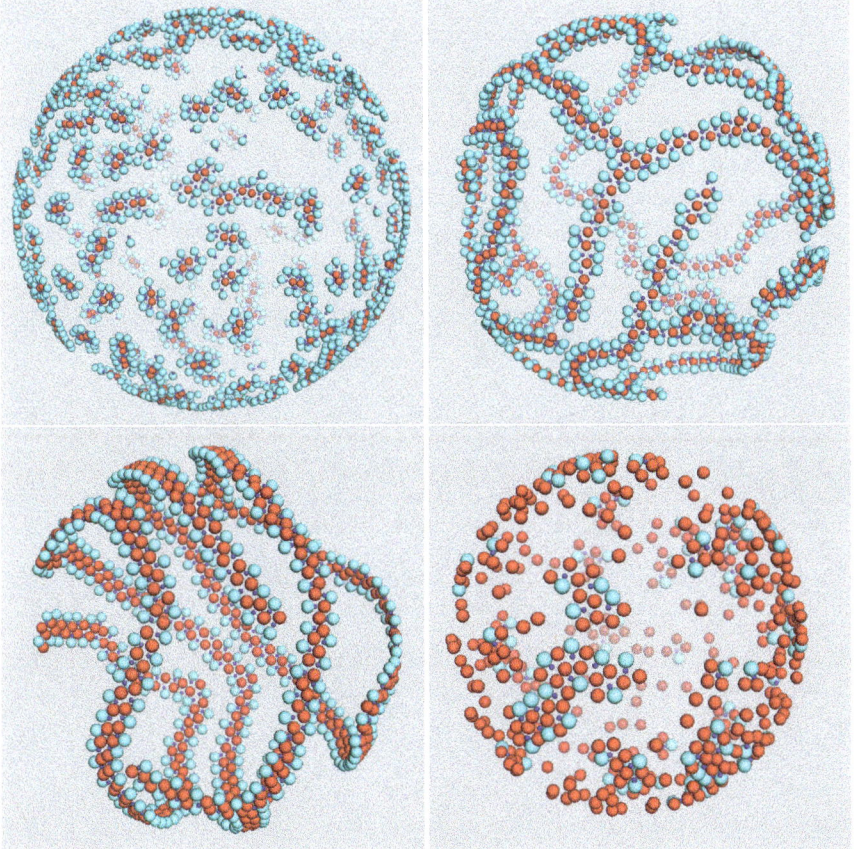

Figure 6. Typical configuration of mixture for $\sigma_3 = \sigma_2$ and $\rho^* = 0.25$ after a long run carried out at $T^* = 0.10$. Snapshots were taken at different compositions: (from top left to bottom right) $\chi = 20\%, 33\%, 50\%$, and $\chi = 80\%$.

While a further, moderate increase in density at fixed $N \approx 1000$ would not change much in the above results, in a huge-density mixture, the formation of self-organized aggregates could encounter much difficulty due to increased relaxation time and surface overcrowding.

3.2. Increasing Disk Size

When the size of disks became sufficiently large, four at least, the large monomers failed to adequately screen the attraction between two disk aggregates, and, at low temperature, we observed the formation of a condensate over the sphere in agreement with what was found for the system absorbed on a flat interface [31]. This is clearly seen in the snapshots reported in Figure 7, which correspond to a typical late-time configuration of the system for $T^* = 0.10, \rho^* = 0.05$, and $\chi = 20\%$. For $\sigma_3/\sigma_2 = 5$, disks mostly occurred in the form of a square-symmetric polycrystalline structure, held together by dimers interspersed between the disks (we counted an average of four dimers in each square center, in accordance with the overall disk composition). The prevalent square order of the system was transparent in the profile of $g_{33}(r)$, where the second and third peaks fall at distances that are in a ratio of $\sqrt{2}$ and 2, respectively, with the location of the first peak. Figure 7 evidently shows that the size of patches with a clear square motif was nonetheless limited, and the reason for this is twofold. First, this may have been the result of an incomplete equilibration; relaxation to equilibrium (coarsening) is slower for a crystallizing system in which many particles are hosted in a single cell. As a result, the spontaneous elimination of crystalline defects takes much longer than it would in a one-component crystal of spherical particles. Even though the onset of crystalline order is favored on a sphere by the lowering of the nucleation barrier [36], this effect was seemingly small at the probed densities where the curvature of the sphere was also small. On the other hand, perfect square order is inherently frustrated on the sphere, and this placed an upper threshold on the size of the ordered patches (this would still be consistent with the existence of a superstructure of patches, akin to the icosahedral superstructure found in dense systems of hard disks on a sphere [37], but we have no evidence for that).

The regular structure observed in Figure 7 finds a correspondence in the crystalline order of DNA-hybridized polystyrene colloids on the surface of oil droplets, as shown in the fluorescence micrographs of [25]. At variance with these triangular-ordered crystalline patches, however, our system is unique in providing a square-symmetric scaffold for the absorption of foreign particles on a spherical substrate.

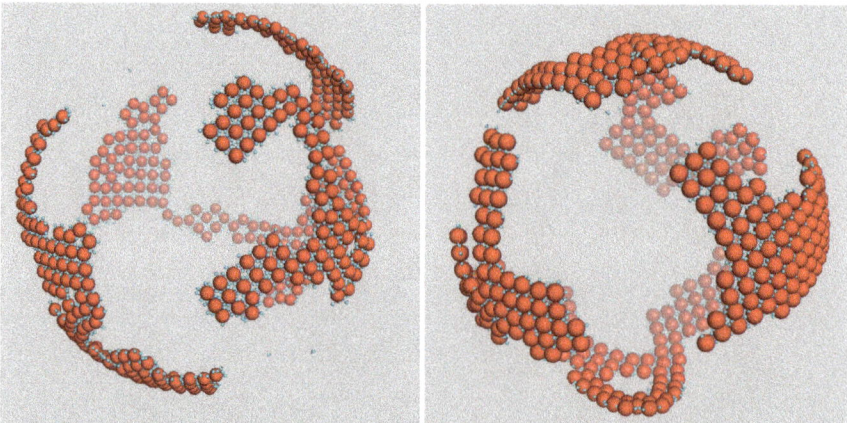

Figure 7. Typical configuration of mixture for (**left**) $\sigma_3/\sigma_2 = 4$ and (**right**) $\sigma_3/\sigma_2 = 5$ after a long run at $T^* = 0.10$. System density was $\rho^* = 0.05$ and disk composition was $\chi = 20\%$. Disks were gathered together in patches similar to tectonic plates floating on Earth's mantle. Prevailing structural motif was a square of disks with gluing dimers in the middle interstice.

4. Conclusions

We performed Monte Carlo (MC) computer simulations on a sphere of a binary mixture of asymmetric dimers of tangent hard disks and guest particles at low temperature. Guest particles were curved hard disks interacting with the smaller monomer of the dimer through an attractive square-well potential, of which the range was the same as the monomer diameter. We analyzed the effect of changing the density of the mixture (while keeping the system very sparse), the composition, and the size of guest particles, ranging from the size of the larger monomer in a dimer to five times larger than that. Despite the simplicity of the model, its self-assembly behavior was quite rich. Indeed, we observed the formation of various metastable aggregates with a prevailing one-dimensional geometry (i.e., chainlike aggregates with a small bending modulus, including a gel-like network), only driven by the short-range attraction, while the large-scale distribution of aggregates could be understood in terms of entropic considerations. For a large guest particle, the small monomer–guest particle attraction could only hardly be screened by the large monomer, thus favoring the formation of thick condensates. These condensates appeared as a square-symmetric polycrystal, which is a nontrivial feature of the model, considering that the perfect square lattice is inherently frustrated on the sphere. Our results indicate the possibility of building a square-symmetric scaffold for the absorption of external particles on a curved surface.

Author Contributions: Investigation, N.D.; Methodology, S.P.; Supervision, G.P.; Writing—review & editing, G.P. and S.P. All authors have read and agreed to the published version of the manuscript.

Funding: This research received no external funding.

Data Availability Statement: Data are available on request from the corresponding author.

Acknowledgments: This work benefited from the computer facilities at the South African Center for High-Performance Computing (CHPC) under allocation MATS0887, and from computer facilities at UKZN (cluster HIPPO).

Conflicts of Interest: The authors declare no conflict of interest.

References

1. Babic, F. *Suspensions of Colloidal Particles and Aggregates*; Springer: Berlin/Heidelberg, Germany, 2016.
2. Weitz, D.A.; Oliveria, M. Fractal Structures Formed by Kinetic Aggregation of Aqueous Gold Colloids. *Phys. Rev. Lett.* **1984**, *52*, 1433–1436. [CrossRef]
3. Lin, M.Y.; Lindsay, H.M.; Weitz, D.A.; Ball, R.C.; Klein, R.; Meakin, P. Universality in colloid aggregation. *Nature* **1989**, *339*, 360–362. [CrossRef]
4. Lin, M.Y.; Lindsay, H.M.; Weitz, ; D.A.; Klein, R.; Ball, R.C., Meakin, P. Universal diffusion-limited colloid aggregation. *J. Phys. Condens. Matter* **1990**, *2*, 3093–3113. [CrossRef]
5. Gonzalez, A.E.; Martinez-Lopez, F.; Moncho-Jord, A.; Hidalgo-Alvarez, R. Simulations of colloidal aggregation with short- and medium-range interactions. *Phys. A* **2004**, *333*, 257–268. [CrossRef]
6. Xu, Z.; Wang, L.; Fang, F.; Fu, Y.; Yin, Z. A Review on Colloidal Self-Assembly and their Applications. *Curr. Nanosci.* **2016**, *12*, 725–746. [CrossRef]
7. Yang, Y.; Pei, H.; Chen, G.; Webb, K.T.; Martinez-Miranda, L.J.; Lloyd, I.K.; Lu, Z.; Liu, K.; Nie, Z. Phase behaviors of colloidal analogs of bent-core liquid crystals. *Sci. Adv.* **2018**, *4*, eaas8829. [CrossRef]
8. Hagan, M.F.; Grason, G.M. Equilibrium mechanisms of self-limiting assembly. *arXiv* **2007**, arXiv:2007.01927.
9. Prestipino, S.; Gazzillo, D.; Munaò, G.; Costa, D. Complex Self-Assembly from Simple Interaction Rules in Model Colloidal Mixtures. *J. Phys. Chem. B* **2019**, *123*, 9272–9280. [CrossRef]
10. Boles, M.A.; Engel, M.; Talapin, D.V. Self-Assembly of Colloidal Nanocrystals: From Intricate Structures to Functional Materials. *Chem. Rev.* **2016**, *116*, 11220–11289. [CrossRef]
11. Thapar, V.; Hanrath, T.; Escobedo, F.A. Entropic self-assembly of freely rotating polyhedral particles confined to a flat interface. *Soft Matter* **2015**, *11*, 1481–1491. [CrossRef]
12. Ramsden, W. Separation of solids in the surface-layers of solutions and suspensions. *Proc. R. Soc. London* **1904**, *72*, 156–164.
13. Pickering, S.U. Emulsions. *J. Chem. Soc. Trans.* **1907**, *91*, 2001–2021. [CrossRef]
14. Evans, D.F.; Wennerström, H. *The Colloidal Domain: Where Physics, Chemistry, Biology, and Technology Meet*, 2nd ed.; Wiley-VCH: Hoboken, NJ, USA, 1999.
15. Morrison, I.D.; Ross, S. *Colloidal Dispersions: Suspensions, Emulsions, and Foams*; Wiley-Interscience: Hoboken, NJ, USA, 2002.

16. Pieranski, P. Two-Dimensional Interfacial Colloidal Crystals. *Phys. Rev. Lett.* **1980**, *45*, 569–572. [CrossRef]
17. Hurd, A.; Schaefer, D.W. Diffusion-Limited Aggregation in Two Dimensions. *Phys. Rev. Lett.* **1985**, *54*, 1043–1046. [CrossRef] [PubMed]
18. Onoda, G.Y. Direct observation of two-dimensional, dynamic clustering and ordering with colloids. *Phys. Rev. Lett.* **1985**, *55*, 226–229. [CrossRef]
19. Ruan, W.D.; Lu, Z.C.; Ji, N.; Wang, C.X.; Zhao, B.; Zhang, J.H. Facile Fabrication of Large Area Polystyrene Colloidal Crystal Monolayer via Surfactant-free Langmuir-Blodgett Technique. *Chem. Res. Chin. Univ.* **2007**, *23*, 712–714. [CrossRef]
20. Retsch, M.; Zhou, Z.; Rivera, S.; Kappl, M.; Zhao, X.S.; Jonas, U.; Li, Q. Fabrication of Large-Area, Transferable Colloidal Monolayers Utilizing Self-Assembly at the Air/Water Interface. *Macromol. Chem. Phys.* **2009**, *210*, 230–241. [CrossRef]
21. Bowick, M.J.; Giomi, L. Two-dimensional matter: order, curvature and defects. *Adv. Phys.* **2009**, *58*, 449–563. [CrossRef]
22. Prestipino, S.; Giaquinta, P.V. Ground state of weakly repulsive soft-core bosons on a sphere. *Phys. Rev. A* **2019**, *99*, 063619. [CrossRef]
23. Tononi, A.; Cinti, F.; Salasnich, L. Quantum Bubbles in Microgravity. *Phys. Rev. Lett.* **2020**, *125*, 010402. [CrossRef]
24. Sacanna, S.; Kegel, W.K.; Philipse, A.P. Thermodynamically Stable Pickering Emulsions. *Phys. Rev. Lett.* **2007**, *98*, 158301. [CrossRef] [PubMed]
25. Joshi, D.; Bargteil, D.; Caciagl, A. Kinetic control of the coverage of oil droplets by DNA-functionalized colloids. *Sci. Adv.* **2016**, *2*, e1600881. [CrossRef] [PubMed]
26. Velev, O.D.; Furusawa, K.; Nagayama, K. Assembly of Latex Particles by Using Emulsion Droplets as Templates. 1. Microstructured Hollow Spheres. *Langmuir* **1996**, *12*, 2374–2384. [CrossRef]
27. Dinsmore, A.D.; Hsu, M.; Nikolaides, M.G.; Marquez, M.; Bausch, A.R.; Weitz, D.A. Colloidosomes: Selectively Permeable Capsules Composed of Colloidal Particles. *Science* **2002**, *298*, 1006–1009. [CrossRef] [PubMed]
28. Munaò, G.; Costa, D.; Prestipino, S.; Caccamo, C. Encapsulation of spherical nanoparticles by colloidal dimers. *Phys. Chem. Chem. Phys.* **2016**, *18*, 24922–24930. [CrossRef] [PubMed]
29. Bott, M.C.; Brader, J.M. Phase separation on the sphere: Patchy particles and self-assembly. *Phys. Rev. E* **2016**, *94*, 012603. [CrossRef] [PubMed]
30. Prestipino, S.; Munaò, G.; Costa, D.; Caccamo, C. Self-assembly in a model colloidal mixture of dimers and spherical particles. *J. Chem. Phys.* **2017**, *146*, 084902. [CrossRef] [PubMed]
31. Prestipino, S.; Munaò, G.; Costa, D.; Pellicane, G.; Caccamo, C. Two-dimensional mixture of amphiphilic dimers and spheres: Self-assembly behaviour. *J. Chem. Phys.* **2017**, *147*, 144902. [CrossRef] [PubMed]
32. Munaò, G.; Costa, D.; Prestipino, S.; Caccamo, C. Aggregation of colloidal spheres mediated by Janus dimers: A Monte Carlo study. *Colloids Surf. A* **2017**, *532*, 397–404. [CrossRef]
33. Krauth, W. *Statistical Mechanics: Algorithms and Computations*; Oxford University Press: Oxford, UK, 2006.
34. Fiumara, G.; Pandaram, O.D.; Pellicane, G.; Saija, F. Theoretical and computer simulation study of phase coexistence of nonadditive hard-disk mixtures. *J. Chem. Phys.* **2014**, *141*, 214508.
35. Pękalski, J.; Ciach, A. Orientational ordering of lamellar structures on closed surfaces. *J. Chem. Phys.* **2018**, *148*, 174902. [CrossRef] [PubMed]
36. Gomez, L.R.; Garcia, N.A.; Vitelli, V.; Lorenzana, J.; Vega, D.A. Phase nucleation in curved space. *Nat. Commun.* **2015**, *6*, 6856. [CrossRef] [PubMed]
37. Prestipino, S.; Ferrario, M.; Giaquinta, P.V. Statistical geometry of hard particles on a sphere: analysis of defects at high density. *Phys. A* **1993**, *201*, 649–665.

MDPI
St. Alban-Anlage 66
4052 Basel
Switzerland
Tel. +41 61 683 77 34
Fax +41 61 302 89 18
www.mdpi.com

Entropy Editorial Office
E-mail: entropy@mdpi.com
www.mdpi.com/journal/entropy